Grey Peacock Pheasant hen incubating eggs

A Birdkeeper's Guide to

Breeding Birds

Pagoda Mynah with leg ring for sexing

The American Robin must have livefood to rear its chicks

A Birdkeeper's Guide to

Breeding Birds

Comprehensive adivice on breeding, rearing and
exhibiting a wide selection of cage and aviary birds

David Alderton

Tetra⬤Press

16083

A Salamander Book

© 1988 Salamander Books Ltd.,
Published in the USA by Tetra Press,
3001 Commerce Street,
Blacksburg, VA 24060.

ISBN 1 56465 159 2

This revised edition © 1997 Salamander Books Ltd.

All correspondence concerning the content of this volume
should be addressed to Tetra Press.

Eggs of the Black-winged Stilt

Credits

Editor: Vera Rogers   Designer: Rod Ferring
Colour reproductions:
Contemporary Lithoplates Ltd..
Filmset: SX Composing DTP
Printed in China

Author

David Alderton has kept and bred a wide variety of birds for more than thirty years. He has travelled extensively in pursuit of this interest, visiting other enthusiasts in various parts of the world, including the United States, Canada and Australia. He has previously written a number of books on avicultural subjects, and contributes regularly to general and specialist publications in the UK and overseas. David studied veterinary medicine at Cambridge University, and now, in addition to writing, runs a highly respected international service that offers advice on the needs of animals kept in both domestic and commercial environments. He is also a Council Member of the Avicultural Society.

Photographer

Cyril Laubscher has been interested in aviculture and ornithology for more than forty years and has travelled extensively in Europe, Australia and Southern Africa photographing wildlife. When he left England for Australia in 1966 as an enthusiastic aviculturalist, this fascination found expression as he began to portray birds photographically. In Australia he met the well-known aviculturalist Stan Sindel and, as a result of this association, seventeen of Cyril's photographs were published in Joseph Forshaw's original book on Australian Parrots in 1969. Since then, his photographs have met with considerable acclaim and the majority of those that appear here were taken specially for this book.

Contents

Page 10–11: Indian Ring-necked Parakeets

Introduction

Interest in breeding birds, as distinct from simply keeping them as decorative aviary occupants or pets, has grown dramatically during recent years. For example, nearly 90% of the parrot species known to be kept in aviaries have produced chicks.

Breeding birds has gained momentum as birdkeepers have acquired a greater understanding of the reproductive needs of their stock, and significant technological advances have accelerated the trend. These advances include the formulation of special diets and improvements in veterinary anaesthesia, the latter enabling surgical sexing to be carried out more safely and reliably. Developments in artificial insemination and incubation have also helped to increase breeding successes, and eggs are often hatched artificially and the chicks hand reared. These methods have been developed principally for use with widely kept species, but can be applied with equal effect to rare and endangered

birds, hopefully raising their reproductive potential. Some breeders have converted their hobby to a business, establishing commercial breeding units for birds; it is obviously more cost effective to cater for a number of chicks at a time, rather than for the offspring from a single clutch.

As greater numbers of a species are bred in collections, so the likelihood of a colour mutation or 'sport' is increased. Recently, there has been a rapid rise in the number of colour mutations and varieties of popular species, such as lovebirds and cockatiels. Some of the rarer colours are very expensive, but by applying the genetic principles of colour inheritance and making careful pairings, it is possible to maximize the likelihood of producing mutation offspring at relatively low cost. This aspect of breeding birds is explained in the following pages, along with the genetic concepts involved in developing a stud of exhibition birds.

Accommodation for breeding

Providing birds with suitable accommodation plays a vital part in encouraging them to breed. In this chapter, we consider how to provide the best housing conditions, either in an outside aviary or in a birdroom.

Aviary breeding
Although some species will nest successfully in the confines of a cage others, such as many softbills, prefer the seclusion of planted aviaries, where they can find natural livefood in the form of invertebrates to rear their chicks. A well-planted aviary is also important for species that tend to become aggressive from the onset of the breeding period. Male pheasants, for example, and other birds, such as touracos, housed in sparse, open aviaries may persecute and even kill prospective partners if there is little cover for the female to retreat to when pursued by an overtly amorous cock bird.

Siting the aviary
Although some birds become tamer when breeding, you should disturb them as little as possible during this period, especially if they have not nested successfully in your collection before. Try to ensure that the birds have as much privacy as possible; if you want the birds to breed outside, site the aviary in a quiet part of the garden, choosing a spot away from the children's play area. Position the flight so that it is not visible from the road, since the headlamps of passing vehicles may disturb sitting hens after dark and cause them to desert their nests. It is important to exclude vermin, such as rats and mice, from the aviary for the same reason and also because they can spread disease to the birds.

Excluding pests
Mount the aviary structure on secure foundations that extend for at least 30cm(12in) below ground level. This helps to discourage burrowing rodents and provides a

Above: *A sense of security in the aviary is vital for nesting birds, such as this Laughing Thrush.*

solid base for the framework. Above ground, clad the flight panels using wire mesh with an individual weave pattern not exceeding 2.5x1.25cm(1x0.5in), which will exclude all but the smallest rodents. You could use mesh measuring 1.25x1.25cm(0.5x0.5in) but this will add considerably to the cost of the aviary. Take similar precautions around waterfowl enclosures, by setting fencing below ground level and making it fox-proof above. It can be a good idea to house smaller ducks in an aviary-type enclosure, since predatory birds, such as magpies, can steal eggs and chicks, given the opportunity.

Access to the aviary
The arrangement of doors in the aviary is an important consideration if the birds are to have a sense of security. The typical aviary is accessible from both the outside flight and the inner protected area, known as the shelter. There are three good reasons for entering the aviary via the shelter. The first is that many species prefer to nest under cover in the outside flight and will be disturbed if you walk through the flight once or twice a day to feed

them. Such disruption can lead to chilling of the eggs. Entering the shelter directly causes little disturbance when the birds are nesting in the flight.

Secondly, entering via the shelter affords you protection, since some larger parrots, especially, may become aggressive during the breeding period. They will not hesitate to attack you, should you approach too close to the nesting box. If they prove particularly aggressive you can confine them in the flight by means of a sliding door over the birds' entrance hole to the shelter.

The third reason is that it is better to feed birds in the shelter, where their food will remain dry. It is also easier to clean up perishable foods spilt on the floor if you line this area with newspaper. A connecting door leads from the shelter to the flight and externally a safety porch surrounding the shelter will prevent any birds escaping as you enter or leave.

Protecting the flight

A layer of translucent plastic sheeting on the roof will prevent nests from becoming saturated during periods of heavy rain and protects recently fledged chicks from becoming saturated. You can also cover the sides of the aviary with plastic sheeting to protect it from rain and draughts and to prevent cats from climbing up the wire mesh, which can cause birds to desert their nests.

The importance of plants

Aviaries for parrots are generally sparsely furnished, because of the birds' destructive capabilities. However, the outside flights of aviaries housing finches and most softbills feature plants. Your choice will be influenced to some extent by the local climate, but try to include a variety of plants with different growing habits. The birds then have a choice of nesting sites within the vegetation.

Conifers, with their fairly dense pattern of growth, are favoured as a breeding retreat by many small species, but be sure to select a slow growing variety that will not rapidly exceed the height of the aviary. Another useful plant in this category is the Box shrub (*Buxus*). Holly (*Ilex aquifolium*) may also provide berries, especially favoured by members of the thrush family, which often choose to nest within its branches. The main disadvantage of this plant is its prickly leaves, although it does remain attractive throughout the whole of the year.

You can include various creepers to provide seclusion, but they, too, may need regular trimming if their growth is too profuse. The popular (but very vigorous) creeper Russian Vine (*Polygonum baldschuanicum*) has the added bonus of colourful flowers that attract insects to the aviary to supplement the birds' regular diet.

Ground cover is especially important in aviaries housing species that nest close to the ground, from the Quail-finch (*Ortygospiza atricollis*) to pheasants and quails. Bamboo (*Sinarudinaria* species), carefully set in clumps, provides a central area of retreat. Replant bamboo at intervals, as clumps die off after a number of years. Golden Rod (*Solidago* hybrids) and Michaelmas Daisy (*Aster novae*) are also valuable plants for this purpose, and are usually easy to grow. Divide these perennial flowering plants as necessary to encourage vigorous growth.

In a small flight it is difficult to maintain a grass floor in good condition, as the earth tends to become waterlogged and moss soon starts to replace the grass. Under these conditions, therefore, you may prefer to plant the vegetation in pots, and stand these on a concrete base that is easy to keep clean. However, you should plant grass in an aviary for pheasants and other birds that spend most of their time on the ground, since constant contact with an abrasive concrete surface can cause foot problems.

Nestboxes in the outside flight

Providing a variety of nesting sites generally increases the likelihood of birds breeding within the aviary.

Always be sure to fix nestboxes securely within the flight. One method is to run a strip of wood down the back of the nestbox and screw this top and bottom to an upright in the flight. Alternatively, you can use brackets or, where space is limited, you may need to fix one or two horizontal lengths of timber outside the flight, running from one vertical support to another, at a suitable height. You can then fix a number of nestboxes along the length of timber through the aviary mesh. For birds bred on a colony system, such as budgerigars, this method provides an effective means of supplying a choice of nesting sites all at the same height. It avoids the likelihood of fighting for the upper nestboxes, as often happens when

A planted aviary

A budgie nestbox, also popular with larger finches

Provide sufficient perches and ensure that they – and any nestboxes attached to them – are secure

Tree nesters, such as barbets, may favour a converted hollow log

A nesting basket firmly suspended

Clumps of bamboo provide height and variety

Slow-growing conifers give cover all year

Trailing nasturtiums add colour and attract livefood

Ground-nesting softbills require plenty of cover in which to conceal their nests

Wash old tree branches and remove leaves to avoid introducing disease into the aviary

Vigorous climbers and creepers need regular trimming, or they may damage aviary mesh

Safety porch

The dense box shrub offers a breeding retreat and berries for thrushes

Paving stones set beneath the perches are easy to clean

A grass floor is important for ground-dwelling birds

Above: *A mixed collection of finches sharing an aviary. Note the wide range of nesting containers under cover in this flight.*

pairs of birds are bred together.

Finch nestboxes are usually made of thin 6mm(0.25in) plywood, and you can arrange these in a similar fashion. However, in an aviary housing a mixed collection of finches, it is a good idea to provide nestboxes at different heights to encourage breeding. The Quail-finch, for example, nests close to the ground. The nestbox may take the form of a small parrot nestbox or it may be open-fronted, in which case there is a raised lip on the box at the front instead of an entrance hole. Wicker nesting baskets are also suitable for finches in aviary surroundings and are easy to fix to the aviary mesh by means of the wire strands at the back of the baskets. If you wish, you can camouflage the baskets with suitable vegetation, such as conifer branches. Alternatively, you can attach the basket securely to a suitable tree or shrub in the aviary.

Hollow logs are quite easy to convert into nestboxes and can prove particularly useful in tempting reticent pairs of softbills to nest. With some notable exceptions, such as barbets, these birds tend not to be destructive. Only choose logs that appear robust and free from obvious fungal contamination, and pay particular attention to the base. If in doubt, cut and fit an artificial floor, so there is no risk of the

15

Nesting sites

Wicker nesting basket for finches

Open-fronted nestbox

Canary nesting pan with felt

Converted hollow log

Budgie nestbox with concave

Above: *A nestbox for the Mandarin Duck, which is tree-dwelling by nature and nests off the ground.*

Left: *A range of nesting sites for breeding birds in cage or aviary surroundings. All but the hollow log are available commercially.*

bottom falling out once the birds have started nesting. Some trees are more suitable than others; silver birch is a popular choice in Europe because of its relatively straight trunk. Avoid elm, however, since moving elm wood from one area to another may also spread Dutch Elm Disease.

The birdroom

A birdroom offers more flexibility for breeding birds than a straightforward aviary. It need not be an elaborate structure and often forms part of the aviary, with one end of the birdroom partitioned off as the shelter for the outside flight. In a birdroom you can incorporate breeding cages and provide artificial heat and light, thus enabling you to extend the breeding period. Most serious exhibition breeders are dependent on a birdroom. It enables them to control pairings that are likely to produce better quality offspring, rather than allow individual birds to choose their own mates.

Breeding cages for small birds

Breeding cages for budgerigars, canaries and finches are easy to obtain, either from the larger pet

Above: *Breeding cages with nestboxes attached both inside and outside. One cage houses Gouldian finches with their young.*

stores or specialist aviculturist suppliers who advertise in the various birdkeeping magazines. Instead of individual cages, some breeders prefer to use so-called 'double' or 'treble' breeding cages. They are obviously more cumbersome than a single cage but can be more versatile, since you can convert them to more spacious stock cages by removing the sliding partitions between the individual units.

Stack the breeding cages in tiers, raised about 30cm(12in) off the floor. Arranged like this they are easier to clean and, furthermore, results are often disappointing if cages stand directly on the floor.

You can build the basic box structure of your own breeding cages, and buy the cage fronts separately. These are available in various sizes, and wire spacing, depending on the group of birds concerned. Relatively large cages are preferable for breeding, so avoid the smaller cage fronts. The extra space created will give the birds more opportunity for exercise, thus keeping them fitter and improving fertility. It is quite possible to construct cages for finches and canaries using oil-tempered hardboard, with the shiny surface forming the interior. Left plain, this surface is easy to wipe with a damp cloth. Many breeders prefer to paint the inner surface of their breeding cages with a light coloured emulsion. Although you can build a budgerigar breeding cage from hardboard, plywood is more durable and less susceptible to damage by the birds' beaks.

Fitting the cage front is straightforward, using timber cut to the appropriate size. Drill holes in the timber to correspond to the horizontal projections on the top and bottom of the front. Then screw the mounted front into the sides of the box. Incorporate a sliding tray, fitted with a handle and raised edges, on the base of the cage. The raised edges will contain the majority of the seedhusks and the handle will enable you to remove the tray for easy cleaning. Although you can line the cage with bird sand, several sheets of newspaper are more absorbent and easier to change. Avoid using coloured sheets on the floor of budgerigar breeding cages, however, in case they contain potentially poisonous inks. Hen budgerigars, in particular, prove destructive when breeding, and often chew up the lining paper on their tray.

Nestboxes in the breeding cage

The position of the nesting site in the cage depends partly upon the species of bird being kept and the choice of nesting sites available. With canaries, it is usual to supply a nesting pan, usually made of plastic, held in place on the back wall of the cage by a screw. Be sure to allow for the height of the adult birds when fixing the nestpan; some breeds, such as the Yorkshire Fancy, are taller than others, and may benefit from a slightly higher cage front from the outset. Without adequate space above them, the birds will be forced to hunch over, which is considered a bad fault in exhibition stock, and they may even find it difficult to feed their chicks.

Smaller finches, such as Zebra and Bengalese (Society) finches, tend to prefer more secluded nesting sites, but relatives of the canary, such as the singing finches, may be tempted to use an open nestpan. Attach the wicker nesting baskets for finches, by their strands of wire, either to the cage front, or with netting staples driven a short distance into the back of the cage. The second option is usually preferable, since it will allow you to look inside the nest more easily. Otherwise the entrance to the basket will face inside the cage and be totally obscured from view.

Other alternatives are small finch nesting boxes, with a partially opened front or a circular entrance hole. Fit them to the back or side of the breeding cage with a small bracket, or even a screw. If you glue them in place, they are more difficult to remove.

Budgerigar nestboxes are considerably larger, and tend to occupy a disproportionate amount of space within a breeding cage. For this reason they are usually attached to the outside of the cage, often on the end. In this case, you will need to cut a hole in the side wall of the cage corresponding to the entrance of the nestbox, and fit an access perch beneath the entrance hole

inside the cage. Support the box on the outside with a suitable bracket. Alternatively, you can screw the box in place, but this is less satisfactory since there is always the risk that the sharp ends of the screws will extend into the

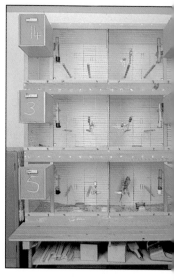

Above: *A tier of breeding cages housing budgerigars. Note the storage area beneath the cages.*

Below: *Budgerigars will breed successfully, using just a wooden concave for their eggs.*

interior of the nestbox and injure the birds and their chicks.

Where space is limited, it is possible to cut away some of the bars at the top right hand corner of the breeding cage (viewed from the front), and site the nestbox in this position. Be sure to shield the budgerigars from the cut edges of the bars by filing these down until they are level with the supporting horizontal bar beneath, or by shielding them with a wooden block. The major drawback of this method is that it permanently damages the cage front, making it more difficult to convert the cage into a flight area outside the breeding period. In the case of a nestbox attached to the end of the cage, simply remove the box and place a wooden flap over the hole. Effectively covering a hole at the front of the cage is far more difficult, especially as the budgerigars will gnaw at any exposed edges of wood that are accessible to them.

Artificial lighting

While most birds tend to nest during the warmer part of the year, some species prove less adaptable in this regard than others. Madagascar lovebirds (*Agapornis cana*) and some members of the parrot genus, *Poicephalus* for example, may show a distinct preference for breeding in the winter months in northern climates. Other birds may adjust their breeding cycle, probably because of a greater response to photoperiodism (i.e. to the relative lengths of day and night). Exposure to light has important effects on the whole body, registering via the hypothalamic portion of the brain. Increasing the daylength by providing artificial light can prove a trigger for breeding activity (see *Conditioning factors* page 30).

The more prolific species, such as budgerigars and cockatiels, will generally continue nesting throughout the year; their reproductive cycles are only clearly depressed during the moulting period. However, the provision of artificial light can benefit them as well. Firstly, it provides them with a longer feeding period, which can in itself act as a conditioning factor. Secondly, the spectral output of a 'natural' fluorescent tube matches that of sunlight and so these tubes are particularly valuable for breeding stock. The birds may be able to synthesise Vitamin D_3, normally produced by sunlight falling on their plumage. Seed tends to be low in this vitamin, which is vital for moving calcium stores around the body, and therefore particularly important during the breeding period. Indeed, a shortage may give rise to soft-shelled eggs, and cause egg-binding in breeding hens (page 41).

With the wide range of electrical equipment available today, it is possible to maintain the lighting in the birdroom under automatic control. Time switches suitable for use with fluorescent lights are advertised in the specialist avicultural magazines. They switch the lights on and off automatically at pre-set times, which can be especially useful on winter nights, when darkness may already have fallen before you return home from work, for example. The most flexible lighting control system relies on a photo-electric cell, located by a window, which responds to the prevailing natural light conditions. As the level of illumination falls, so the lights in the birdroom are automatically switched on; they are turned off as the outside light level improves.

A dimming device in the lighting circuitry is recommended, so that the birdroom is not plunged immediately into total darkness when the lights are turned off. This could leave the birds stranded outside their nestbox, causing eggs or chicks to become chilled. The benefits of a dimmer are also apparent in the morning; the lights can be gradually switched on before you enter the birdroom, even though it is still dark outside. Suddenly turning on the lights could disturb sitting birds.

Heating

Heating the birdroom is often a contentious issue among birdkeepers, but if the birds are breeding during the colder part of the year, you must certainly provide adequate additional warmth. Without this, eggs are more likely to be fatally chilled when the adults leave the nest, and chicks are rendered more vulnerable to the effects of cold. Furthermore, the additional energy used to maintain body heat may affect the young birds' growth.

Although you can fit gas and paraffin (kerosene) heaters in a birdroom, they have proved dangerous and each year birds die as a result of accidents involving such heaters. Electricity is undoubtedly the safest means of heating a birdroom, but avoid fan heaters, since they tend to clog with dust after a short period in operation. Tubular heaters with a central element are suitable, however, since they are sealed units and impervious to dust. For maximum efficiency, mount them off the floor so that air can circulate around them and thus permit the transfer of heat. If necessary, you can arrange two or more tubular heaters above one another on the wall. The number needed will be influenced by the size of the building and the degree of insulation, as well as by the temperature required. Ideally, ensure that the internal temperature does not fall below 7°C(45°F) when the birds are breeding. Although birds may lay satisfactorily in a colder environment, there is an increased risk of egg-binding occurring under these conditions.

Various thermostats operate in conjunction with tubular heaters; sometimes these plug into the power socket, and the plug on the heater then simply slots into the thermostat. Check the range over which the thermostat operates – most are effective between 0-30°C(32-86°F) – and the wattage which it will control. It may also be possible to incorporate an alarm system, should the unit fail for any reason. This is an especially valuable precaution with breeding stock, as it may save both eggs and chicks during cold weather.

In a large area, you may choose to use a portable convector heater with an output of several kilowatts, compared with the lower wattage of the tubular heaters. Portable convector heaters usually incorporate a thermostat to control heat output, depending on the external environmental temperature. They may also feature a programmable timer, so the heater can be pre-set to switch on and off at specific times.

Ionizers

Another piece of electrical equipment – the ionizer – has proved valuable for improving the environment within a birdroom and assisting with breeding results. When switched on, the electrical current produces a constant stream of electrons from the needle tip of the ionizer. These collide with molecules in the air to form negative ions, that 'cluster' around dust particles, airborne bacteria and other microorganisms. Having acquired a negative charge, these particles are precipitated from the atmosphere to an earthed surface, such as the floor, and can be wiped up. An ionizer is also capable of destroying disease-causing particles directly, as well as making the atmosphere cleaner.

Below: *An ionizer will help to ensure a cleaner atmosphere within a birdroom, which should lead to better breeding results.*

Nestboxes suitable for parrots

Species	Width	Height	Depth
Lovebirds	20cm(8in)	25cm(10in)	20cm(8in)
Cockatiel Grass parakeets	25cm(10in)	30cm(12in)	25cm(10in)
Many lories Pionus parrots	25cm(10in)	50cm(20in)	25cm(10in)
Amazons Grey parrots	30cm(12in)	50cm(20in)	30cm(12in)
Smaller cockatoos, e.g. Goffin's	38cm(15in)	50cm(20in)	38cm(15in)
Larger cockatoos, e.g. Moluccan	45cm(18in)	50cm(20in)	45cm(18in)
Large Macaws e.g. Blue and Gold	75cm(30in)	90cm(36in)	75cm(30in)

Breeding parrots indoors

In some parts of the world, such as northern Canada, the climate precludes the keeping of tropical birds in outside aviaries throughout the year. Furthermore, many parrots, especially the large species, are noisy and it is not always possible to keep them in an outside aviary in an urban area. To overcome these problems, you can use a separate birdroom to house these birds, or even convert a spare room in the home. Good insulation, in the form of special quilting or polystyrene placed behind the walls and roof of the birdroom, will help to muffle the raucous calls of these parrots. Special insulation board may also be useful for this purpose. You can use plywood or oil-tempered hardboard to line the birdroom and to conceal the insulation material.

Within the birdroom, you can house parrots in all-wire flight cages. These are relatively easy to make, using wire mesh panels held together with special metal clips. It is best to use large open mesh, perhaps 5x2.5cm(2x1in), depending on the species concerned, to form the floor of the flight cage. Parrot droppings tend to be rather tenacious, but they will fall through the mesh quite easily, along with seed husks and discarded fruit. For the sides and top of the cage, use 2.5cm square (1in square) mesh. Sheets of newspaper underneath the flight cage are easy to remove, so you can keep the birds' surroundings quite clean, with relatively little interference at breeding time.

Secure the cage using suitable hooks fitted to the side of the birdroom, or raise it up on brick pillars at each corner, with an extra course of bricks to retain the four corners in place. The location and size of the flight within the birdroom will determine the best means of fixing it in place. Locate the service door into the flight cage so that the interior is easily accessible, and you can replace perches without difficulty.

Parrot nestboxes

You can buy suitable parrot nestboxes either in kit form or ready-made, but they tend to be expensive, especially in terms of carriage, because of the weight of the timber used in their construction. Most psittacines need a nestbox constructed from timber with a minimum thickness of 2.5cm (1in); the actual dimensions of the box will vary according to the species concerned. While the smallest species, such as parrotlets, may readily adopt a budgerigar nestbox, most require more spacious nesting surroundings. The table gives an indication of the relative sizes of breeding cage for different psittacines.

It is usually easier to position the nestbox outside, attached to one end of the flight. Secure it firmly so there is no risk of it becoming dislodged. By cutting away just sufficient mesh to give access to the nestbox opening, it is possible to shield most of the wooden front from the parrots' beaks.

Parrots become particularly destructive when they are in breeding condition, and will rapidly whittle away any accessible wood incorporated into the nestbox. The entrance hole is clearly vulnerable to further expansion by their beaks, but if too much light is allowed to penetrate the interior of the nestbox, the birds may not feel sufficiently secure to use this site for breeding purposes, and thus fail to nest. For this reason, some breeders cover the edges of the entrance hole with metal sheeting, which is knocked flat and held in place by small nails. It is easier to protect a nestbox with a square entrance hole than one with the more traditional circular access.

If you decide to construct the nestbox yourself you will also find it easier to cut a square entrance. When planning the box, remember that in order to check on the birds during the breeding period, you will need a suitable inspection flap. Indeed, the roof of the nestbox must either lift off completely, or be hinged so that it opens away from you. The most secure method of fixing the roof is to design it as a separate section, so that it is just over 2.5cm(1in) wider than the nestbox. You can then fix a 2.5cm(1in) framework around the sides of the roof, which slots over the top of the nestbox, holding the roof firmly in position.

In addition, you may decide to pre-drill a hole through the 2.5cm(1in) supporting frame, so

that when you assemble the box you can insert a screw to anchor the roof and box firmly together. When you want to lift off the lid, simply undo this screw. With this system there is no need to fit hinges of any kind.

Usually the nestbox hangs close to the roof of the aviary or near the top of a flight cage, so that a hinged lid is less practical than a removable roof. Even so, it can be difficult to gain access by this means, so you may wish to add a side entrance to the nestbox. This is usually located about 15cm(6in) from the floor of the nestbox so that when the flap is opened there is no risk of eggs or chicks falling out. Hinge the flap on the upper edge, and secure it with a hook and eye at each lower corner.

Alternatively, you can incorporate a removable sliding inspection hatch on the side. Make the hatch about 2.5cm(1in) bigger than the hole in the side of the nestbox, so that it does not fall into the box. Attach strips of wood to form runners that hold the hatch in place externally. A handle of some kind will enable you to pull the hatch back and forth without difficulty. The major drawback of a side inspection door is that the parrots will undoubtedly find it easier to destroy the interior of the nestbox, starting along the exposed edges of wood around the hatch.

You may be able to divert their attention from the structure of the nestbox by providing small offcuts of untreated softwood on the floor that they can whittle away to provide a lining for the nestbox. It is clear from observations of wild parrots that they may spend a considerable period of time preparing their nest site, so providing them with wood to chew helps to satisfy a natural instinct. Other nest linings, such as peat, are sometimes recommended, but use these with caution. Peat tends to get very dusty once it dries out and many birds do not like it, frequently scraping it out onto the floor of their surroundings. Rotten

'Grandfather clock' nestbox

Removable roof for access ⸻

Entrance hole ⸻

Internal wire mesh ladder, firmly secured, for access to base ⸻

Perch for easy entry ⸻

Hinged side inspection door

Below: *A macaw using its beak and claws to climb a mesh ladder.*

wood is sometimes used, but can be harmful to the birds' health, since it contains fungal spores that may infect the parrots while they are using the nestbox.

The risk to health from contaminated nest litter has been confirmed by studies carried out in the wild. These have shown that Philippine Red-vented Cockatoos (*Cacatua haematuropygia*) can be badly affected with the fungal disease aspergillosis, caused by mouldy nest litter.

Deep nestboxes

Australian parakeets, especially, tend to favour the deeper design of nestbox, sometimes described as the 'grandfather clock' type. It is a good idea to include a ladder to facilitate access to the bottom of all nestboxes of this type in order to prevent accidental injury to either eggs or chicks. Be sure to fix the ladder firmly in position; if it becomes dislodged, the adult birds will probably be unable to reach their chicks and, sadly, each year whole clutches are lost for this reason.

Fix the ladder in place before assembling the nestbox. Use aviary mesh for the ladder; 16 gauge (16G) is ideal for most species, although thinner 19G can be used in many cases. Cut the mesh carefully so that no loose sharp ends of wire can injure the bird. File down cut ends, and then trim the mesh so that it runs from the bottom of the entrance hole to about 10cm(4in) from the bottom of the nestbox. Tack two thin pieces of battening down the inside of the nestbox as a base for the ladder, and then attach the ladder to the battening using netting staples. Most breeders rely on 'artificial' nestboxes, but for large psittacines, including the large *Ara* macaws, you may simply prefer to convert an old oak cask, cutting an entry hole in one end. As with the more conventional nestbox, you will find it easier to cut a square, rather than a circular, hole. In either case, it will need to be about 15cm(6in) across.

Other nesting receptacles for these birds are less satisfactory, especially converted metal dustbins. The heat within them builds up very rapidly and the birds cannot move in and out so easily. Condensation inside the bin may also become a problem. Hollow logs are sometimes used, but since most psittacines will readily take to an artificial nesting site, there is no real advantage in providing a box of this type.

Sexing Birds

Preparations for the breeding period should begin several months before any eggs are anticipated. Birds take a variable length of time to settle down in new quarters; often, imported psittacines will not attempt to breed for several years after completing their period of quarantine. In the first instance, however, it is important that you identify true pairs. In the past, this presented considerable difficulties, since only a relatively small proportion of psittacines can be sexed by visual means, such as the plumage differences between cocks and hens or the difference in cere coloration in budgerigars. Over the last decade or so, the situation has changed dramatically, and in cases where sexual dimorphism does not occur, it is now possible to determine the gender of such birds with certainty.

Above: *It is easy to sex a pair of adult cockatiels, since the cock has clear yellow face markings.*

Anatomical signs

Physical features, such as the size of the head, may be applicable as a means of sexing in some species. In many of the larger *Ara* macaws, for example, cock birds tend to have a broader forehead, but local variations in size render such means of distinction largely invalid. As cock birds of other species mature, however, they may produce more of the dark pigment, melanin, responsible for black coloration. Young cock cockatiels of the Pearl mutation darken in colour as they mature, for example, although the Cockatiel is sexually dimorphic.

In the Grey Parrot, however, this secondary sexual characteristic can be a help in recognizing pairs, especially if a dealer is reluctant to have the birds sexed. This can happen if a dealer has experienced a serious imbalance between the two sexes in one consignment and has found it more difficult to sell the birds as separate known sexes rather than as unsexed individuals. In the Grey Parrot, mature males tend to have a darker back and wings than hens from the same geographical area, although there can be considerable variation in the depth of coloration between individual birds.

The pelvic bone test

During the last century, scientists looked for methods of sexing monomorphic species (i.e. those in which the males and females are visually identical) by external means. Nearly a third of all bird species fall into this category. They discovered that hen birds in breeding condition have a wider gap between their pelvic bones to permit the easy passage of the eggs. With the bird resting on its back in the palm of the hand, the pelvic bones can be located as two bony prominences either side of the midline, just above the vent. The change at breeding time can be quite noticeable; in *Ducula* fruit pigeons, the gap may accommodate the index finger during the breeding phase, but at other times the space is greatly reduced, and the bony projections almost touch each other.

Since the distinction is effectively lost outside the breeding phase, relying on the so-called 'pelvic bone test' tends to yield an abnormally large number

of believed cocks that eventually prove to be adult hens out of breeding conditions or immatures.

Behavioural signs

Although behaviour can provide a useful clue to the sex of birds, it is far from being a reliable indicator. Indeed, two parrots of the same sex housed together may frequently act as a pair, with one feeding the other, and they may even attempt to copulate. With two hen birds, eggs can result, although these tend to be much greater in number than the usual clutch and are, of course, infertile.

During the breeding period, members of the finch genus *Lonchura*, including the widely kept Bengalese Finch (*Lonchura domestica*), can be sexed by the song of the cock bird, and indeed most birds become far more vocal at this time of year.

Endoscopic sexing

At present, one of the most widely used methods of sexing birds is surgical, or endoscopic, sexing. By inserting an instrument known as an endoscope into the bird's body cavity, it is possible to view the internal organs, including the reproductive system in the abdominal area. This procedure is usually carried out under a general anaesthetic and so, depending to

some extent on the species concerned, it is usually necessary to withhold food from the bird for a period beforehand. Check this with your veterinarian in advance, and also notify him or her if you suspect that your bird is ill in any way or breathing abnormally, as this may affect the use of an anaesthetic. In addition, obesity can cause problems during this procedure. Birds recover far more quickly from a gaseous anaesthetic than they do from an injectable agent.

The small hole required to insert the delicate endoscope is made by an instrument known as a trocar. Before the examination, the veterinarian will remove a small area of feathers and clean the site where the endoscope will be inserted – on the left side of the body behind the last rib. The reason for choosing this site is simply that most hen birds retain only one functional ovary located on their left side. In contrast, cocks usually have two testes, located close to the kidney, one on each side of the vertebral column.

Once the endoscope is in position, the inspection to determine the bird's gender is quick, taking just a minute or so. The endoscopic examination may also reveal problems, such as inflammation and discoloration of the normally transparent air-sacs, thus emphasizing the accompanying diagnostic value of the procedure. The bird can then be treated as required.

After the examination, it is not usual to suture the tiny opening through which the endoscope has been inserted into the body. Indeed, it will be barely visible by the time the bird has recovered consciousness. If possible, keep the bird on its own for a few days so that it can recover fully without being harassed by other birds, depending again on the advice of your veterinarian.

In experienced hands, the mortality associated with surgical sexing is very low. Studies suggest that in percentage terms it is less than 0.5%. Deaths that do occur are often the result of a concurrent disease problem, such as aspergillosis, which may not have been appreciated before the examination took place, and anaesthetic complications. The procedure is safe for all but the smallest birds, and although it is most commonly used in psittacines (parrot family) and raptors (birds of prey), it can also be safely applied to softbills weighing in excess of 30gm (1oz).

Surgical sexing does have its drawbacks as a means of sex determination, most notably the difficulty of determining the gender of young birds. Indeed, it can be difficult to sex immature individuals, since there are no follicles present on the ovary – a feature in an older hen bird that clearly distinguishes it from a male of the same species. Similarly, the testicles of cock birds enlarge dramatically when they are in breeding condition, to such an extent that there is relatively little to see through the endoscope.

DNA sexing

This method of sexing, based on the body's genetic material, has become very popular during the 1990s. It is safer and often more convenient than endoscopic

Below: *The tip of the endoscope (illuminated) entering the trocar inserted into the body cavity.*

sexing. There is no need for the bird to be given an anaesthetic, and the required blood sample can be taken at home. Just a tiny volume is required, from a cut claw or blood feather, and is placed in a special tube, which is then sent direct to the test laboratory.

Another major advantage of this method is that chicks can even be sexed reliably in the nest, although the procedure is normally carried out after fledging.

Accurate early sexing can be of crucial importance to breeders of

Below: *The site of incision for an endoscopic examination. Only a few feathers need to be removed.*

Below: *Viewing the internal sex organs with an endoscope. This is a relatively safe way of sexing birds.*

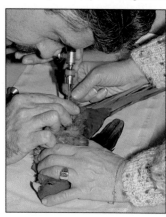

Anatomy of the male bird

The air-sacs form part of the bird's respiratory system, and are normally quite transparent.

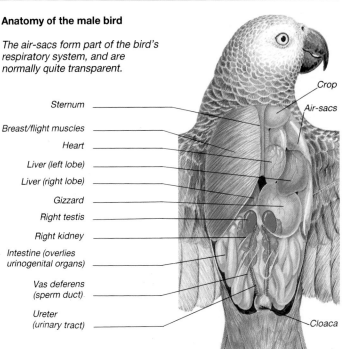

Sternum

Breast/flight muscles

Heart

Liver (left lobe)

Liver (right lobe)

Gizzard

Right testis

Right kidney

Intestine (overlies urinogenital organs)

Vas deferens (sperm duct)

Ureter (urinary tract)

Crop

Air-sacs

Cloaca

Anatomy of the female bird

Usually only the left ovary and oviduct are functional; the right oviduct is rudimentary.

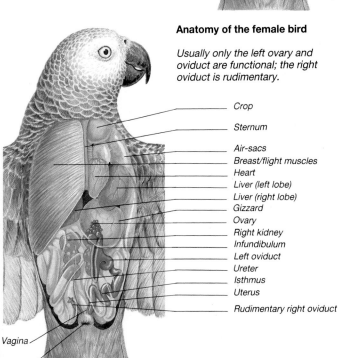

Crop

Sternum

Air-sacs

Breast/flight muscles

Heart

Liver (left lobe)

Liver (right lobe)

Gizzard

Ovary

Right kidney

Infundibulum

Left oviduct

Ureter

Isthmus

Uterus

Rudimentary right oviduct

Vagina

Cloaca

sex-linked recessive colour mutations (see page 82), such as the lutino form of the Ring-necked Parakeet. In this instance, it is important to determine the sex of the chicks as soon as possible, so that decisions about which birds to retain can be taken at an early stage, lessening the pressure on aviary space as a result.

Green cock birds from the mating of a normal green cock to a lutino hen will carry the lutino characteristic in their genetic make-up, and they can in turn produce lutino offspring in the next generation. Hens in the same nest will lack the lutino gene, but it is impossible to distinguish the sexes visually at this stage. DNA sexing overcomes this difficulty.

Similarly, it can also be very helpful where young cockatoos are concerned. Pairing young cockatoo chicks from an early age is likely to lead to fewer problems of compatibility once the birds are mature. Introducing two adult cockatoos can be fraught with danger, however, as the cock may attack his intended mate.

Chromosomal study
Whereas DNA sexing focuses on the genetic material, this method relies on a study of the chromosomes, which are present within the nucleus of each living cell in the body. One pair, called the sex chromosomes, are responsible for determining the bird's gender. It is possible to distinguish the sexes by the relative difference in length between the sex chromosomes of male and female birds. Here the situation is the reverse of that in mammals; the hen rather than the cock has one chromosome shorter than the other. This is the so-called ZY combination, with 'Y' indicating the shorter chromosome. The cock bird has a ZZ pairing.

This method of sex determination, called chromosomal karyotyping, came to prominence in the 1980s, but it has since been superseded as the main laboratory method by safe and convenient DNA analysis.

Steroidal study
While both endoscopic and DNA sexing have grown in popularity during recent years, the original laboratory method – faecal steroid analysis – has tended to be overlooked. In the early 1970s it appeared to be of potentially great importance, but its acceptance was hindered because it was not a precise method, relying instead upon a ratio.

In terms of hormone balance, females produce a larger quantity of oestrogen, whereas males produce more testosterone. By measuring the relative amounts of these chemicals present in the bird's faeces, it is theoretically possible to determine the gender of an individual bird. Where there is a high ratio of oestrogen to testosterone, i.e. in excess of 2.5:1 the bird is likely to be a female. If the ratio favours testosterone, then this indicates a cock bird.

Unfortunately, the relative hormonal output is influenced by the maturity of the bird in question and its reproductive state. Only mature and reproductively active birds give consistent results when this technique is applied. However, if sequential samples are taken and analyzed for their relative steroidal content, the method can be used to show when birds are coming into breeding condition.

The research surrounding faecal steroid analysis confirmed that the relative ratio of oestrogen to testosterone varied between individual species, so that it was not possible to establish a common baseline. Indeed, to check the results obtained when birds have been grouped on the basis of faecal steroid analysis, it proved necessary to sex them surgically as well. Clearly, breeders were not keen to pay twice to have their birds sexed and, as could be expected, the most reliable method became more popular, even though faecal steroid analysis was a totally safe method of determining the sex of the birds. Research in this field has continued with newly hatched chicks.

Certification

If you buy a pair of parrots that have been sexed by surgical or laboratory means, you may well be presented with a certificate showing when and where the examination was carried out. It is not unusual for a breeder to cut a small distinguishing mark in the tail feather of either cocks or hens, so that these birds can be distinguished on a temporary basis. You may prefer to have them marked permanently with a tattoo, usually applied under the wing. Tattooing is also used as a security measure for more valuable stock, but it may fade in time and the markings become illegible. As a method of distinguishing pairs of birds, where only one member of a pair is tattooed, it forms a reliable guide to identification.

Alternatively, you may need to resort to banding. Split celluloid rings can be used for softbills, but for larger parrots, special stainless steel rings are a safer option (see pages 56-7). Micro-chipping is another method of identification that can be used. The microchip unit, about the size of a rice grain, is implanted under the skin, and can only be detected by means of a special reader, which activates the concealed chip.

Below: *A Java Sparrow with two rings: closed metal for age and origins, split plastic for sexing.*

Conditioning factors and compatibility

A number of different factors influence the likelihood of breeding success, and the breeding triggers themselves are clearly complex. Only when all the conditions are correct is a pair likely to nest successfully. Although considerable efforts have been made during recent years to improve the chances of hatching and rearing chicks successfully, relatively little attention has been given to studying the stimuli that trigger reproductive activity in the first place. Once this aspect of breeding is better understood, it may become possible to persuade reticent pairs of birds to nest.

It is clear that the part of the brain known as the hypothalamus is responsible for triggering reproductive activity, and that hormones ultimately stimulate the development of the reproductive organs in both sexes. Various factors may stimulate the hypothalamus to begin the chain of activity that ultimately results in eggs being laid. Here we consider the most important factors.

Daylength

The importance of increasing daylength has already been mentioned (see page 19). You can control daylength by housing birds inside, and altering their exposure to light over a period of time.

Photoperiodism (i.e. the relative lengths of day and night) tends to be the major determining force in the reproductive cycles of species that originate from more temperate areas, away from the equator. For them, the warmer part of the year is clearly the most favourable time for nesting. The canary, unlike the budgerigar, does not breed throughout the year for this reason. Indeed, the nesting period tends to be quite rigorously defined in the case of these birds.

While a progressive increase in daylength is generally seen as the trigger for breeding, the situation may be reversed. Certain birds are known to nest in response to a falling photoperiod, when daylength is declining. The most significant avicultural example is probably the Ne-ne or Hawaiian Goose (*Branta sandvicensis*), a species which has been saved from probable extinction by captive breeding. The Emu (*Dromaius novaehollandiae*) is an Australian example of a species that actually nests during the winter months.

Below: *An indoor flight within a birdroom, complete with electric light. Exposure to light can be a conditioning factor for breeding.*

Water

In species that originate from arid areas, such as budgerigars and many other Australian parakeets, nesting may continue as long as environmental factors are favourable. As might be expected, rainfall appears to act as a breeding trigger to such species. This means that as long as they have plenty of water, as well as suitable food, such birds are usually free-breeding in aviary surroundings. However, a shortage of water and a fall in temperature may prove to be inhibiting factors.

The sound of running water, such as a waterfall in a tropical house, may be a stimulus for breeding in various seedeaters and softbills. You can, of course, provide artificial bathing facilities for birds housed indoors, using either plant sprayers or a hose and fine nozzle for this purpose. Although this may not directly encourage breeding activity in some species, it will help to maintain their overall condition.

Diet

Many of the species kept in bird collections originate from equatorial parts of the world, where daylength tends to be fairly constant throughout the year. In these cases the impact of photoperiodism on the birds' reproductive cycle is normally minimal in the wild, and it seems likely that the birds achieve control of their breeding cycle by internal means. They nest on a regular basis and breeding is stimulated by a recurring environmental trigger, perhaps a change in the available foods. In softbills, for example, an increase in the numbers of insects could be a significant factor.

It is clear that, assuming other conditions are satisfactory, many birds can be encouraged to nest by raising the protein level of their diet. Parrot breeders feed their birds on a diet that contains more protein than non-breeding rations, with a particular emphasis on the essential amino-acids, which cannot by synthesized in the body. For the same reason it is a good idea to offer birds rearing food before the start of the breeding period. It will help to compensate for any shortcomings in the existing diet and means that by the time the chicks hatch the parent birds should be taking the rearing food readily.

Below: *Diet, especially livefood, plays an important part in stimulating breeding behaviour and is vital for successful rearing.*

Nesting material

The provision of suitable nesting sites within the birds' environment and an adequate supply of nesting material can have a significant effect on the birds' willingness to breed. Offer the birds nesting facilities at least a month or so before you hope they will produce fertile eggs. Where nestboxes have been provided for roosting purposes throughout the year, give them a thorough spray with a suitable aerosol to kill any red mite parasites. Apart from taking this precaution, all you need supply is appropriate nesting material.

Pet stores sell proprietary nesting material, suitable for most seedeaters and softbills, in the form of paper strips. Other items, such as dried grass and moss, should be available from a florist. Avoid using hay, which is frequently mouldy after storage. Larger softbills and pigeons will use a selection of twigs to form their nest, but ensure that these have been cut to the correct size for the birds concerned. Provide a platform of some kind, particularly for pigeons, as they tend to be sloppy nest-builders when left to their own devices. Without support, eggs and chicks may be lost.

Parrots, unlike most other birds, tend not to use nesting material of this type, preferring to rely instead on a layer of dry wood chippings on the floor of the nestbox (see page 22). Budgerigars will simply lay and rear their chicks on a wooden concave, designed to fit the nestbox. Notable exceptions, however, are the lovebirds (*Agapornis* species) and the hanging parrots (*Loriculus* species), both of which actively collect twigs and leaves in order to line their nesting chamber. Whereas the Abyssinian Lovebird (*Agapornis taranta*) builds a simple pad on the floor of the nestbox, members of the so-called 'white eye-ring' group, such as Fischer's Lovebird (*A. fischeri*), tend to weave a far more ornate nest, building a domed structure within the nestbox. They carry nesting

Sterile nest lining

Twigs

Dried grass

Leaves

Shredded paper

Coconut fibre

String-like nesting material

material in their beaks, whereas the Peach-faced Lovebird (*A. roseicollis*) and hanging parrots tuck suitable lengths into the plumage at the rear of the body and rump, which enables them to transport more at a time.

Pair bonding

This important phenomenon may also have a direct bearing on reproductive success. Some birds remain with a single mate throughout their lives, forming a permanent pair bond. Such behaviour is common in larger waterfowl, notably swans and geese. In other cases the bond may be only temporary, with the birds remaining together through the winter and then splitting up in the spring, after egg laying has occurred. Mallard ducks (*Anas platyrhynchos*) are one example.

In some cases, such as budgerigars, the birds normally form a strong pair bond and cage breeding may tend to compromise fertility if an established pair in the aviary are split up for breeding purposes. To overcome this problem, some budgie breeders house cocks and hens in separate aviaries throughout the year.

Among parrots, the larger species, especially, seem to form permanent pair bonds. If one member of the pair dies, it can be

difficult to persuade the survivor to accept another mate, and in fact this may take several years. Dominance is another factor to consider when pairing up certain psittacines, notably those belonging to the genus *Psittacula*. In these parakeets, hens are dominant to cocks for most of the year, except during the breeding period. Mutual preening – a common sign of bonding in many members of the family – is not seen in these parakeets. Indeed, outside the breeding period, the male may be actively persecuted by his partner if he approaches too near to her.

As the time for nesting becomes close, however, the hen will permit the cock bird to advance, so that mating can take place. If possible, do not pair a young, inexperienced cock Ring-necked Parakeet (*Psittacula krameri*) or Alexandrine Parakeet (*P. eupatria*), for example, to a mature hen, since he may be too scared of his partner to mate with her. Ideally, therefore, the reverse pairing is recommended.

Behavioural changes

As well as nestbuilding, birds will show other signs of their readiness to breed. Some species of weaver and whydah, for example, moult into so-called 'nuptial plumage' just before the onset of the breeding period, and may be described as being 'I.F.C.' (In Full Colour). For the remainder of the year they are much duller and tend to resemble the hen, being 'O.O.C.' (Out Of Colour). Many waterfowl also have duller, so-called 'eclipse' plumage outside the breeding period.

We have already seen that it is possible to sex some species at breeding time because the cock bird becomes much more vocal. The increasing frequency of the cock canary's song, for example, also serves to stimulate successful nesting. A further function of the male bird's song is to advertise his presence in the territory and warn off other individuals that he considers may be 'trespassing'.

Below: *Canaries tending a chick on a nestpan. Breeders often use felt to line the nest site.*

Aggression at breeding time

Aggression becomes more apparent from the onset of the breeding period and, in aviary surroundings, birds that have lived together in harmony may lose their docile natures, especially towards related species. For example, pairs of blue waxbills, also known as cordon bleus, may become aggressive towards each other, although they are more tolerant of waxbills of different coloration. If possible, accommodate individual pairs of these waxbills in small planted aviaries.

It may be necessary to remove some individuals from a communal aviary if they are harassed by other birds, otherwise physical harm may result. Unpaired birds and recently introduced individuals are most at risk. Avoid outbreaks of fighting in a communal aviary by ensuring that there is adequate cover for the occupants and do not overcrowd the enclosure – the best results are likely to occur if the birds breed unmolested.

In the case of finches that normally nest in groups, such as the Java Sparrow and other related *Lonchura* species, establish the colony before the start of the breeding period so that the birds have an opportunity to settle down and establish a flock order. In these cases, breeding results will certainly be better if a number of birds of the same species are housed together. Even so, just one or two pairs may breed, rather than the whole group. By spacing the nesting sites widely around the aviary, rather than concentrating them in one area, you may encourage more pairs to nest.

The generally free-breeding Budgerigar often fails to nest if one pair is housed in isolation. These gregarious psittacines breed more readily when kept within sight and sound of others of their kind.

Dealing with aggression

Aggression during the breeding period need not be confined to birds of the same sex. Indeed, some cocks can become

exceptionally savage towards their intended mates, and may even kill them. In these cases it is likely that the cock attained breeding condition before the hen and became frustrated by her failure to accept his advances. There is little that can be done if the hen comes into breeding condition before her intended partner as mating is then less likely to take place. Individuals do vary considerably in this regard, and compatibility can be a significant factor. Generally, quarrels are less likely between the members of a pair that have bred satisfactorily in the past, rather than between two birds that have recently been introduced.

Some Australian parakeets can be very savage towards intended mates, so it is a good idea to pair them up in the autumn after one breeding period so that the birds can grow accustomed to each other before the next. It can be dangerous to introduce even a mature hen alongside a cock in the spring, when breeding is likely to be imminent. If you doubt the temperament of your bird, remove the cock from the aviary before

introducing the hen and leave her to settle down for several days. Watch the pair for signs of aggression when you replace the cock - a weekend when you are at home is a good time to do this.

You may need to clip one wing of the cock to restrict his mobility, thus enabling the hen to escape if she is seriously threatened. This is not a difficult task, but it will be easier if someone is available to help you. Open one wing and use a sharp pair of scissors to cut in a straight line across the primary and secondary flight feathers, leaving

Above: *In a mixed aviary, fighting may occur during the breeding season. Certain birds may need to be moved elsewhere.*

the outermost two and the innermost three intact. Avoid cutting below the bottom of the vanes, because the shaft of newly emerged feathers receives a blood supply and if you cut them too short you will cause bleeding. Clipping the wing will not handicap the parakeet on a permanent basis, since the feathers will be replaced at the next moult.

Wing clipping

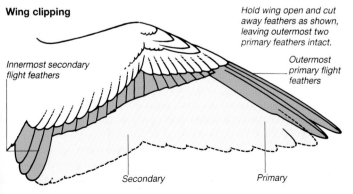

Hold wing open and cut away feathers as shown, leaving outermost two primary feathers intact.

Innermost secondary flight feathers

Outermost primary flight feathers

Secondary

Primary

The laying period

When the hen is ready to mate, she will usually indicate her willingness by flattening herself down into a more horizontal position, enabling the cock to step onto her back. In New World psittacines, however, this situation is slightly altered, as the cock bird keeps one foot on the perch and grips the hen with the other. The cloacal openings are held together, to facilitate transfer of the spermatozoa. Birds lack the penile organ present in mammals, although a rudimentary swelling of this type is present in the cloaca of various male waterfowl and some other species.

The egg

The development of the egg starts in the ovary of the hen. Each ovum that is released is covered in an elastic membrane that protects it during its passage through the upper part of the muscular oviduct. Fertilization takes place close to the top of the oviduct in the funnel-shaped infundibulum, and then the ovum passes into the magnum, which is the longest portion of the oviduct. At this stage, the white albumen is added over a period of about three hours.

From here, the ovum enters the isthmus, where the shell membranes are produced. After about an hour in this section of the oviduct, the egg moves into the so-called 'shell gland'. This section receives a good blood supply, and calcium is removed from the circulation at this point to form the eggshell. It may take nearly a day before this stage of the egg's development is completed.

Finally, the egg is moved into the vagina – the terminal part of the oviduct – and enters directly into the cloaca. From here the egg is

Development of the egg
The diagram at right shows the hen's reproductive tract and where the major stages in the development of the egg occur. The internal regulation of reproductive activity is under the control of chemical messengers called hormones. These are released from the pituitary gland in the brain and circulated around the body in the bloodstream, exerting their effect when they reach their target organs. As with mammals, there is a mass of immature ova within the adult hen's ovary, only a small proportion of which will develop during her life. Involuntary muscular contraction moves the developing egg through the tract.

Internal structure of an egg
The main elements within an egg are shown at right. Shell colour is added in the shell gland, using pigments derived from the breakdown of blood cells. Hole-nesting species have white eggs, but those nesting in the open use colour to conceal eggs from predators. The percentage of yolk in the egg varies; ducks tend to have the highest quantity of yolk.

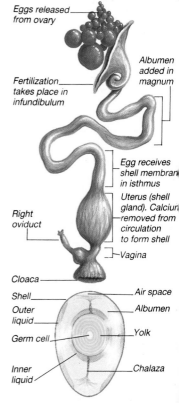

Eggs released from ovary

Albumen added in magnum

Fertilization takes place in infundibulum

Egg receives shell membrane in isthmus

Uterus (shell gland). Calcium removed from circulation to form shell

Right oviduct

Vagina

Cloaca

Shell

Outer liquid

Germ cell

Inner liquid

Air space

Albumen

Yolk

Chalaza

finally laid as the strong muscular walls of the vagina force the egg out of the reproductive tract. The whole process, from fertilization to the fully formed egg reaching the cloaca, takes about a day, although eggs that complete their passage through to the vagina at dusk are not usually laid until the following morning.

Egg laying

Immediately before laying her eggs, the hen will spend longer periods on her nest, or within the nestbox and is often joined by the cock. A slight swelling of the vent area suggests that laying is imminent; avoid handling the birds unnecessarily at this stage. The hen's droppings may also alter, becoming larger and smellier.

Only one successful mating is required to fertilize a clutch of eggs. Most birds can be grouped as 'determinate' layers, meaning that they lay a finite number of eggs in one clutch. However, bantams and other fowl will continue laying on a regular basis as their eggs are removed.

The frequency of laying depends on the species concerned. Finches, for example, lay every day, but parrots generally produce their eggs on alternate days. This means that there may be a large gap between the ages of the youngest and oldest chicks in a nest. As a means of reducing this period – and thus increasing the likelihood of survival of all the chicks – many birds do not start incubating the eggs until two or three have been laid. You may think that the birds are not going to sit properly, if they are frequently off the nest even though there are eggs within. Do not be concerned, however, until the clutch is virtually complete. Avoid disturbing the birds in any way, since they may desert their eggs before incubation begins in earnest.

In order to prevent further development until the clutch is complete, most canary breeders remove the eggs in sequence as they are laid. They replace the eggs with special dummy eggs made of plastic. This is common practice, and makes it easy to predict when hatching will take place. The fertile eggs are carefully stored in a cool place until the morning of the fourth day. They are then replaced under the hen, and the dummy eggs removed. This completes the usual clutch of four and all the eggs should hatch on the same day, ensuring that the chicks are more likely to survive.

Above: *Storing budgie eggs. When the clutch is complete they will be returned to the nest.*

Artificial insemination

At present, this technique is most commonly applied to turkeys and pheasants, but it has also proved successful with parrots. A sample of sperm is obtained from the male and transferred directly into the hen's oviduct. It can be of value with rare birds that are incompatible for any reason, or in cases where a cock bird is unable to mate normally, because of a fatty tumour (lipoma), for example. Your local veterinarian can advise you, or put you in touch with a colleague who is experienced in using this technique with poultry.

Schematic development of an egg from laying to point of hatching

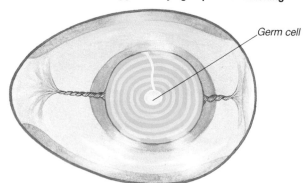

Germ cell

1 In a newly laid egg, a white spot is clearly visible on the yolk. Within this area is the female germ cell and, if the egg has been successfully fertilized, then this part will divide to form the embryo.

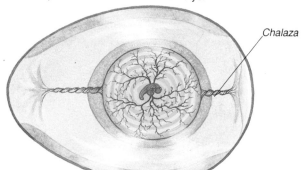

Chalaza

2 Here the developing embryo is clearly visible. Support is provided by spiral bands of dense albumen, the chalazae, which hold the yolk in place when the egg is turned – an essential procedure throughout incubation.

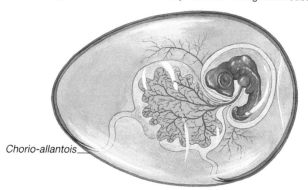

Chorio-allantois

3 As it grows, the embryo needs more oxygen and, similarly, it must prevent an early build-up of carbon dioxide inside the shell. As a result, the chorion and allantois fuse together, forming the chorio-allantois.

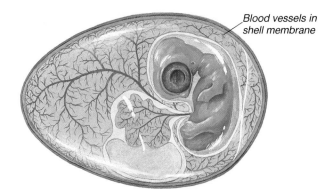

Blood vessels in shell membrane

4 The chorio-allantois quickly forms a lining around the inside of the shell, establishing a mesh of blood vessels here. This acts like a lung, enabling gaseous exchange to take place through minute shell pores.

Air space

5 The orientation of the embryo begins to alter as the time for hatching approaches, so it can break into the air space at the broader end of the shell. The yolk is increasingly absorbed into the embryo's body.

Egg-tooth

6 Just before hatching, the remains of the yolk sac disappear, and the increase in carbon dioxide causes the chick's head to jerk, breaking through into the air space. Its egg-tooth then cuts through the shell.

Incubating the eggs

In most parrots, the hen alone is responsible for incubating the eggs, although the cock frequently spends periods of time alongside her in the nest. If a hen parrot dies, or falls seriously ill while there are eggs in the nest, remove the eggs for artificial incubation or foster them to another pair, since the cock bird will ignore them, even in the absence of the hen.

Cockatoos and cockatiels generally share the incubation. Cock birds tend to sit during the day, with hens incubating for the remainder of the time. Canaries do not share the incubation duties, but many finches will take turns at sitting, as do most softbills and pigeons. Only in phalaropes (wading birds), which are not popular avicultural subjects, does the cock bird sit alone.

In cases where both birds incubate, it is possible for one member of the pair to continue sitting and hatch the eggs alone if an accident befalls their partner. Nevertheless, the chances of successful breeding are reduced under these circumstances.

The desire to incubate is under hormonal control, and both sexes can be affected accordingly. Heat is transferred from the bird's body to the eggs via the so-called brood patches. Down feathers are lost on the underside of the body and the skin here receives a high blood supply, providing local heat for incubation of the eggs.

Below: *A Canada Goose incubating her eggs. Some birds, especially waterfowl, can become quite aggressive when nesting and resent any disturbance.*

Egg-binding

Keep a close watch on all laying hens to ensure that they do not succumb to egg-binding. A lack of calcium and cold, damp weather around the egg laying period may increase the likelihood of egg-binding. The symptoms are virtually unmistakable. You will see the hen emerge from her nest, often in the morning when an egg is expected, appearing weak and reluctant to fly. Her ability to perch is affected, and you may see her on the floor of the aviary or cage.

Handle her gently, since it is important not to break the egg causing the obstruction in her body. An examination of the vent usually shows that it is rather inflamed and swollen, and contractions of the muscles here may be apparent. With the hen restrained on her back, gently feel above the vent and you should locate the presence of an egg.

This is an emergency situation and you will need to deal with the problem at once in the hope of saving the hen. Transfer the bird to a warm place and maintain the temperature at 27°C (81°F) by means of an infrared lamp. The warmth may stimulate egg laying within a few hours; if it does not, you must take more radical action. It may be possible to manipulate the egg carefully from the vagina and out of the body via the cloaca. First apply a suitable lubricant, such as olive oil, into the vent, using a 5ml syringe without a needle attached, and then gently massage the egg free.

The most effective and less stressful treatment for egg-binding, especially in psittacines, is an injection of calcium borogluconate given by your veterinarian into the breast muscle, at a dose of 0.5mg per 100g (3.5oz) body weight. The success of this technique suggests that a deficiency of free calcium in the blood is the underlying trigger for the condition in many cases. It could be that the egg has an incomplete or soft shell, or that the lack of calcium serves to depress muscle contractions. In the latter case, the muscles forming the wall of the vagina are unable to operate effectively, and thus the egg is retained here.

As long as the bird is handled gently, an injection is a relatively safe method of dealing with egg-binding, since there is no risk of breaking the egg within the bird's body. If it should rupture, there is always a danger that an internal infection – often described as peritonitis – may follow, with fatal consequences.

In a particularly stubborn case of egg-binding, your veterinarian may be forced to operate in order to save the bird. A full recovery is possible under these circumstances, and the bird may even breed again in future. Always allow a suitable period to elapse, however, before encouraging further egg laying. It is generally best to wait until the following year before pairing the bird up again.

You can help to prevent egg-binding by providing your birds with a suitable diet that contains adequate calcium. Parrots and other seedeaters should always have access to cuttlefish bone. You can buy this from your seed supplier, and secure it in the cage or aviary with a special clip. The birds will gnaw at the softer powdery side, although smaller finches find it easier to nibble at small slivers cut off with a sharp knife. There are also soluble calcium preparations, which can be given in the drinking water, but take care to avoid overdosing.

In obese birds, immatures or old individuals the reproductive tract may not be functioning effectively, and this makes them more susceptible to egg-binding. Breeding birds out of season, when the temperature is low, may increase the incidence of this problem, even in healthy stock, since chilling alone will depress muscular activity. Although abnormally large eggs may be responsible for egg-binding, they are a rare cause, compared with the other contributory factors.

Prolapse of the oviduct

A prolapse of the oviduct is a common complication following egg-binding, especially if the hen had been straining for a long period before the egg was released. The sign of a prolapse is unmistakable – you will see a variable amount of red, inflamed tissue hanging out of the vent. The tissue is part of the oviduct, usually the vaginal component, but may include the cloaca in many cases. Although it looks very unpleasant, a prolapse appears to cause the bird no great discomfort. You must deal with the condition promptly, however, to minimize the risk of a subsequent infection, which can prove fatal.

If someone else holds the bird for you, it should not be too difficult to replace the protruding tissue. Prepare a bowl of tepid water – a clean, empty margarine tub or similar container is ideal for this purpose – and cotton wool. Wash your hands thoroughly and then bathe the tissue as necessary to remove any obvious dirt or debris adhering to it. With the thumb and first finger of one hand on either side of the vent, gently open the orifice. It should then be possible to push the protruding tissue back inside with the index finger of your other hand. Then release the grip with the first hand.

Hopefully, the prolapsed tissue will remain within the body, but in some cases repeated prolapsing occurs. You can try to replace it several times, but if this is unsuccessful, contact your veterinarian, who will place a so-called 'purse-string suture' around the vent to keep the prolapse from reappearing. After a few days, the problem should resolve itself and the suture can be removed. The situation will be complicated, however, if another egg is anticipated. In many cases, egg laying may be terminated by egg-binding and a resulting prolapse, but it can still occur. There is little you can do, but watch closely to ensure a further prolapse, if it occurs, is dealt with at once.

Other egg problems

Eggs without proper shells are frequently associated with cases of egg-binding, and may result from a problem with calcium metabolism (see page 41). However, soft-shelled eggs may be associated with a separate disorder known as salpingitis, sometimes described as impaction of the oviduct. This is often accompanied by a slight staining of the plumage around the vent, although staining may also indicate a digestive disorder. Salpingitis may result in several soft-shelled eggs being laid in succession.

A variety of bacteria have been isolated in cases of salpingitis, but the most common is *E. coli*. The bacteria may enter the oviduct via the cloaca, together with foreign bodies, such as the abrasive sand particles used to cover the floor of breeding cages. Parasites can also be introduced to the reproductive tract and lead to salpingitis. Treatment with antibiotics is difficult, and existing damage to the oviduct means that it is unlikely that an affected hen will be able to breed normally again.

In waterfowl and some galliformes (domestic fowl and related birds), fluke infestations of the oviduct can cause soft-shelled eggs. Various chemicals may also affect the thickness of the eggshell. The effects of DDT have been well documented in wild birds of prey, but it is important to take care when administering drugs to laying birds in captivity. For example, certain members of the sulphonamide group can interfere with shell formation.

Occasionally, you may encounter an interesting phenomenon, namely the double-yolked egg. This is perhaps most commonly reported in budgerigars, but is also quite well documented in canaries. It arises when two ova are released almost simultaneously from the ovary and pass through the oviduct together, becoming encased at the same time in the shell gland. Studies show that in the majority of double-yolked

The hospital cage

A sick bird may be more willing to drink from a small sealed container close to the perch.

A soaked millet spray may be easier to eat than hard seed, but remove perishable food regularly.

Position perches close to food, water and the heat source.

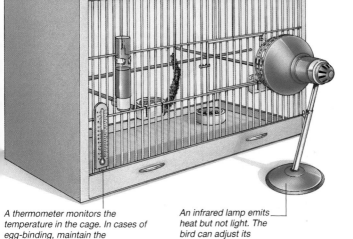

A thermometer monitors the temperature in the cage. In cases of egg-binding, maintain the temperature at about 27°C(81°F).

An infrared lamp emits heat but not light. The bird can adjust its position in relation to it.

budgerigar eggs, the yolks remain separate, although in a third of cases they are fused and could give rise to joined chicks (Siamese twinning). The fertility of double-yolked eggs is low – probably less than 10% – but each year a few do hatch and are reported in the various birdkeeping journals.

Above: *Heat plays a vital part in assisting the recovery of a sick bird. Egg-bound hens may lay normally if moved to a warm place.*

Below: *A purpose-built hospital cage suitable for small birds. Such units allow close monitoring of temperature and lighting levels.*

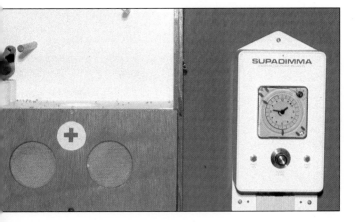

Egg damage

The shell provides the egg with considerable protection, but if it is damaged, then disease-causing microorganisms can gain access to the egg and adversely affect hatchability. Any breaks in the shell will also allow fluid to be lost at a faster rate than normal, threatening the viability of the developing embryo. Eggs can be damaged within the nest by being roughly rolled about or, more often, punctured by the claws of the incubating birds. Budgerigars and canaries tend to be greater offenders in this respect than other species, possibly because they are bred in cages and more frequently disturbed than birds nesting in aviary surroundings.

With budgerigars, always ensure that the wooden concave used to line the nestbox contains a sufficiently deep hollow to retain the eggs, and position this as far from the entrance hole as possible. This should help to prevent the eggs being scattered when the birds move in and out of the nestbox. Never lift the inspection flap without warning, since this will cause the birds to scurry out of the box and increases the likelihood of eggs being damaged. Instead, tap gently on the outside and allow the birds to emerge before you look inside. It is a good idea to train them to accept this procedure before egg laying begins.

Many canary breeders inspect their breeding hens before placing them in the breeding cages, since overgrown claws may puncture the eggs. It is a simple matter to trim

the claws back with a pair of bone clippers. Scissors are less suitable, since they do not cut cleanly through the tissue and may cause the claw to split. Gently restrain the canary in the palm of your hand and, in a good light, locate the blood supply, visible as a thin red streak running some way down the claw. Cut a short distance away from the point where the streak disappears, so there is no risk of bleeding.

It is possible to patch eggs that have sustained minor damage, and they will hatch uneventfully, producing healthy, normal chicks. Carry out such treatment as soon as you notice any flaws and before the damage has a serious effect. Nail varnish is an ideal means of repair, and it is well worth keeping a bottle in the birdroom for use in emergencies. Ensure that your hands are clean and hold the egg gently but firmly, carefully avoiding the area of damage so that you can paint over the crack. Be sure to cover the whole area, but do not use more nail varnish than necessary, otherwise it will block the pores over the undamaged portion of the shell. Allow the varnish to dry before replacing the egg in the nest, otherwise it may stick to another of the clutch, causing further complications.

Sometimes, birds destroy their eggs deliberately and they are then termed 'egg-eaters'. This vice is perhaps most common in

Below: *Overgrown claws may puncture eggs or drag chicks from the nest. Trim regularly.*

Above: *Minor cracks in the shell may be repaired by painting the affected area with nail varnish.*

pheasants and budgerigars. Although the cause is unclear, it is possible to overcome this unfortunate habit. Usually it is not too difficult to distinguish between an egg that is accidentally broken in the nest and one that has been deliberately destroyed. In the latter instance, you may find beak marks in the remnants of the shell and the offending bird often has yolk stains around its face.

When confronted with this problem in budgerigars, you can construct a false bottom to the nestbox. Cut a hole in the concave large enough to allow the egg to fall through, and then support the concave on blocks on the floor of the nestbox. Arrange a bed of sawdust on the nestbox floor to provide a soft surface onto which

the eggs can roll as they are laid. You can then transfer the eggs to another pair that are at the same stage of producing eggs.

You can often correct the vice by obtaining some dummy eggs and simply placing them under the sitting hen. Once the budgerigars realize that the dummy eggs cannot be destroyed, they should desist from destroying their own eggs in the future. Either or both of the parent birds may be responsible for the problem, but generally this method will cure them most effectively. It may be that their diet is deficient in some respect, so that they resort to eating their eggs. If this is the underlying cause, try offering a softfood to supplement the diet.

A similar method can be used to cure pheasants, but obviously the replacement eggs need to be larger – dummy pigeon eggs may be suitable. The traditional method practised by gamekeepers entails placing an old, infertile egg – ideally laid during the previous breeding season – under the hen. Such eggs will be particularly foul if broken, and their revolting taste is said to act as a reliable deterrent. Another method is to add pepper and mustard through a hole in the shell of a fresh egg, which is then patched. Both methods rely on the bird's sense of taste to dissuade it from eating its own eggs in future. Although birds have far fewer taste buds on their tongues than mammals, they do have some sense of taste, and so such methods can be effective and are worth attempting.

A nestbox with a false base

A hole in the concave allows the eggs to drop out of reach of egg-eating budgerigars. Retrieve the eggs for fostering to other pairs.

A layer of sawdust or other soft substrate protects the eggs as they roll through.

Hatching and rearing

How can you tell if the eggs have hatched? The adult birds remaining in the nestbox after the incubation period has passed and an increase in food intake are early signs that the eggs have hatched. (If the eggs are not going to hatch the parents will generally have lost interest in them by this stage.) You may also see pieces of eggshell in the aviary, often discarded by the adult birds some distance from the nest. Or you may hear the calls of the chicks, often just before dusk, as they solicit food from their parents. There is no need to worry if you do not hear them calling consistently; in fact, this is usually a sign that all is going well. Only if they are cold or short of food will chicks prove consistently vocal.

The hatching process

The young chick cuts its way out of the shell using a so-called 'egg-tooth' located on the upper beak. This is lost soon after the chick has hatched. Occasionally, hatching does not appear to progress normally once the chick has 'pipped' through the shell membranes and made a hole in the outer shell. Do not rush in to help the chick out as soon as you suspect that something is amiss, since you can cause a fatal haemorrhage if you puncture any of the blood vessels in the shell membrane. Once these blood vessels have disappeared the chick is fully ready to hatch.

Check first that the chick has been able to pierce the air space at the larger (blunt) end of the egg, by gently peeling off a piece of shell. Provided it has cut through here safely, the chick will be able to survive, feeding off the remaining yolk sac reserves that have nourished it right through the incubation period. After a further day or so, when the blood vessels have disappeared from the shell membrane, the chick will be ready to hatch. If necessary, a chick can survive for as long as two days after pipping, but do not leave it until it is so weak that it cannot survive once freed from its shell.

Hatching difficulties are quite scarce, but are most common in budgerigars. Chicks hatched artificially in an incubator are probably more at risk than those left with adult birds (see page 69). Contrary to some beliefs, chicks that are helped out of the egg do not normally suffer from a genetic weakness and develop into normal healthy birds.

Help with hatching

When helping a chick to hatch, be sure to start by washing your hands so there is less risk of causing a yolk-sac infection. A pair of blunt-ended tweezers are useful for lifting off pieces of shell and dissecting the shell membranes since you can use the points to lever off pieces of shell through any existing hole, whereas this will be much harder if you have to rely on your fingers.Try to avoid exerting downward pressure on the egg when you are freeing the chick. You can also use tweezers to tear the shell membranes apart, employing a technique known as 'blunt dissection'. A second pair of tweezers and a small pair of scissors will make the task easier. Start by lifting the shell membranes up using one pair of tweezers, and nip a small area with the scissors to create a hole. Lift the membranes at this point with one pair of tweezers, and insert the second pair just into the hole. Gradually allow them to expand, thus enlarging the tear in the membrane. Any blood vessels that remain are more likely to clot if the membranes are torn in this way, rather than being cut, because of the resulting spasm in the tissue. Carry out this procedure in a warm environment, so that the chick does not become chilled.

Once most of the shell has been removed, along with the membranes, the chick should be able to free itself without further difficulty. Before replacing the young bird in the nest, hold it for a few moments in the palm of one hand and cover it gently with the other hand to warm it up with your

body heat. Dab the umbilicus with an iodine solution or a baby wound powder to protect against any infection gaining access to the body and to dry up the area.

Check the chick again several hours later. Hopefully, the parent birds will have fed it by this time. You can confirm this by looking at the crop at the base of the neck. If there is no sign of any food here you may need to feed the chick yourself (see page 71) so that it can gain strength. In a group of strong siblings, weaker offspring are often neglected at first if they do not hold their heads up to receive food. Once they are stronger, these chicks will be fed without problems by their parents.

The rearing period

Chicks can be divided into two groups on hatching. In some bird families, the chicks emerge from the egg blind and helpless and totally dependent on the adult birds for their survival. Passerines, such as finches, as well as parrots, pigeons and birds of prey, are all typical examples of these so-called 'altricial' species. The term originates from the Latin word *altrix*, translated as 'nurse'. In

Above: *A Muscovy Duck hatching from its egg. Having cut around the top of the egg, the duckling forces the cap off and wriggles out of the shell. The remains of the yolk sac will help to sustain the duckling for the first day or so of its life out of the egg. It will be wet on hatching, and not until the plumage has dried out will it have the usual fluffy appearance associated with young ducklings. Later on, its plumage will acquire the necessary waterproofing that allows it to swim without its feathers becoming saturated. Water is vital, but remember that young ducklings can become chilled – or may even drown – at this stage, so do not provide an open bowl of deep water for them.*

contrast, the offspring of waterfowl, pheasants and other game birds are described as 'precocial'. They hatch with a thick covering of down that insulates their bodies when they are not being brooded, and these youngsters are immediately capable of following their parents around and feeding independently, although they may be less able than adult birds to find food.

Since precocial species have a more hazardous early life than altricial birds, their clutch size tends to be relatively larger. It is easier to rear precocial species, however, since the chicks are much more straightforward to look after. They rapidly learn to feed themselves and many are capable of flying within days of hatching.

The fledging time of altricial chicks varies considerably. Some finches may leave the nest within 10 days or so of hatching, although they may not be able to fly properly. At this stage they hide away in the vegetation and are cared for principally by the cock bird. Typical examples are the various buntings in the family Emberizidae. By contrast, it may be several months before the larger parrots leave their nest; macaws, for example, are likely to be about 15 weeks old when they emerge for the first time.

Above: *Altricial chicks, such as budgerigars, cannot fend for themselves and need feeding.*

Below: *Precocial chicks, such as Whooper Swan cygnets, are well developed when they hatch.*

Nest hygiene

Under normal circumstances, there is no need to interfere with the nest once the chicks have hatched. The inside of the nest is kept clean by the adult birds and, in many cases, the droppings of the chicks dry out quite rapidly, forming a powdery base. It is quite normal to clean out budgerigar nestboxes, however, since here the birds are being reared on just a piece of wood, and there is no absorbent nest litter. You can generally leave the nestbox untouched until all the eggs have hatched; the frequency of cleaning depends partly on the number of chicks and also on the build-up of dirt within the nest. Some pairs – often described as 'wet feeders' – have excessively damp interiors to their nestboxes, and these may need cleaning once, or even twice, a day.

The tenacious nature of the droppings means that they can become stuck to the young birds' developing claws. You must remove this encrusted dirt, otherwise permanent malformation of the claw – or even partial loss of the toe – may result and ruin the exhibition potential of a particular chick, even before it has left the nest. Always soak off the dirt; trying to remove this material without first softening it is likely to cause damage to the claw, and possibly even bleeding.

Start by filling a container, such as a clean yoghurt tub, with tepid water and then allow the chick's dirty foot to hang down into it. Gentle handling is essential at all times, since young birds are, of course, delicate creatures. After a minute or so, the debris should have softened sufficiently for you to chip it off gently with your finger nail. If minor bleeding does occur, dab the affected area with a safe germicide ointment and the chick will recover uneventfully.

When removing young budgerigars from their nestbox, you will need another container for them while you change the wooden concave. A small plastic bowl lined with paper towelling is ideal for this purpose. You can dispose of the lining paper as each group of chicks is replaced. You may notice that once the budgerigars are about three weeks old, they will start nibbling on the floor of the nestbox. They may actually eat their droppings, since these provide a source of Vitamin B. The vitamin is synthesized by bacteria present in the lower part of the digestive tract, but since the bird cannot absorb vitamins from this part of its body it must eat the droppings to digest them in a higher region of the gut.

In some instances, however, droppings and food become stuck on the lower surface of the upper beak. Hidden from view, this dirt is likely to lead to malformation of the beak unless it is removed. Using the blunt end of a matchstick (with its head removed) or a cocktail stick, you can generally break away the debris from the beak in a single piece. Less commonly, dirt can also accumulate under the tongue near the lower beak. Treat this in a similar way, to prevent the beak becoming 'undershot'. In this case the lower portion of beak grows abnormally, causing the upper part to curl behind it, rather than overlapping it as is normal.

Generally, domesticated species do not resent limited interference in their breeding activities, but you need to exercise great care with other birds, especially if they have not bred before in your collection. Having nested successfully at least once, it is surprising how much more tolerant pairs can prove to be when they breed again.

Be particularly cautious when inspecting the nests of cockatoos (and indeed other large parrots), since they may attack you. This applies especially to tame birds that have little or no fear of people. Some species, such as the Bare-eyed Cockatoo (*Cacatua sanguinea*), are usually pugnacious in any event. Try to arrange the aviary set-up so that you can feed such birds without going near their nestbox. Close yourself in the shelter if necessary.

Below: *As it feathers up, check the inside of a budgerigar's beak for any deposits of food and nest dirt that may have accumulated here. Clean these off without delay.*

Feeding for rearing

Providing a nutritious diet for the parent birds can be vital in ensuring that chicks are reared successfully. Most finches, for example – normally seedeating birds – become largely insectivorous when they have chicks in the nest. Failure to provide an adequate selection and quantity of invertebrates throughout this phase is likely to lead to the premature death of the chicks. It is important to make plans for the rearing period well in advance, therefore, by setting up cultures of suitable livefoods. Here we consider a selection of livefoods used by birdkeepers.

Whiteworms These tiny worms (up to 1cm/0.4in) long are an ideal soft-bodied food for small softbills, as well as seedeaters. They are easy to raise. Obtain a starter culture and prepare a suitable breeding container, such as a clean, empty margarine tub (or similar). Fill this close to the top with damp peat and moisten some wholemeal bread with milk, to act as a source of food for the whiteworms. Make a series of small shallow holes in the peat, just below the surface, and spoon in the mixture of bread and milk. Then divide up the whiteworm starter culture, placing some of the worms on each food source. Finally cover them with peat and place the lid (complete with ventilation holes) on top. Transfer the container to a warm environment where the temperature can be maintained at about 20°C(68°F).

Make sure that the culture is not allowed to dry out and that the whiteworms are well fed. They should be ready for harvesting within a month; simply separate them from the peat by dropping a spoonful of the medium complete with worms into a saucer of water.

You can culture other, smaller worms of the *Enchytraeus* genus in a similar way; microworms and grindalworms are both valuable foods during the breeding season.

Above: *A plastic container with ventilation holes for culturing whiteworms. Keep the peat damp and regularly change the bread soaked in milk so it does not sour.*

Above: *After a month, separate the worms from the culture medium by dropping them in a dish of water.*

Set up several cultures in succession so that you have sufficient to supply the birds throughout the rearing period.

Fruit flies (*Drosophila*) are a particularly valuable livefood for small softbills. Both hummingbirds and sunbirds need regular supplies of this type of livefood as part of their normal daily diet, and if these birds do decide to nest in your aviary, the quantity of livefood they consume will increase dramatically during the rearing phase. Some breeders set up a culture within the birds' quarters, giving them a virtually free supply of these nutritious invertebrates.

Fruit flies will multiply readily on a variety of foodstuffs, such as old banana skins, but you can now buy specially formulated preparations that are simply mixed with water to act as a breeding medium. These are generally safer to include in the birds' quarters since, unlike decomposing fruit, there is no odour or mould associated with this culture.

Crickets are especially versatile as a livefood. They are normally sold by number, rather than by weight, and are available in a range of sizes from tiny hatchlings to adults measuring about 15mm(0.6in) in length. Keep them in a converted aquarium covered with a muslin or fine mesh cover. Since crickets need warmth to survive, you may need to use one of the special vivarium lids if the temperature of their environment is likely to fall below 24°C(75°F). These lids include space for a light bulb to act as a heat source. The crickets eat a mixture of flour and grass. You can seed the grass in trays so that it need not be cut fresh. Spray the grass each day and the crickets will drink the droplets of water on the shoots. A saturated paper tissue is another means of supplying moisture.

Crickets are quite easy to breed. You can recognize females by the longer abdomen that terminates in the egg laying portion, known as the ovipositor. Provide pots of damp sand where the females can lay their eggs. Keep the surface of the containers moist using a light spray, always ensuring that the contents do not become flooded. The eggs should start hatching about three weeks after they were laid. You can offer young crickets to small seedeaters as well as to softbills, but first cool the crickets down in a refrigerator so they are less likely to escape once released. As the temperature falls, the crickets become less active.

Stick insects are easy to maintain and in their long life cycle produce soft-bodied hatchlings that are an ideal rearing food for birds. Choose the Indian Stick Insect (*Carausius morosus*), since this is a relatively hardy variety that thrives at room temperature. You need not worry about pairing up these invertebrates; females lay fertile eggs on their own (this process is known as parthenogenetic reproduction), and males of the species are exceptionally rare. Although stick insects will feed on

51

privet, do not offer it to hatchlings that are destined to become food for the birds; there is a possibility that privet remaining in the hatchlings' bodies could prove toxic to the birds. Feed the hatchlings on bramble leaves to avoid this risk.

Mealworms are the larval stage in the life cycle of the Meal Beetle (*Tenebrio molitor*). They are readily obtainable, by mail-order if necessary, from the specialist suppliers who advertise in the avicultural magazines. However, they are not suitable for rearing newly hatched chicks, although they are a valuable food for adult stock. They have a hard indigestible covering of chitin, which fledglings are unable to break down in the digestive tract. When they moult, mealworms shed this layer for a period and at this stage they appear white, rather than yellowish brown. In this state, they can be used for rearing purposes. In an emergency, you can scald mealworms to break

Below: A variety of livefoods can be valuable at breeding time.

down the casing and make them safer for the parent birds to offer to their chicks.

Countering deficiencies
Livefood is often deficient in certain essential nutrients. Mealworms, for example, are relatively low in calcium, and chicks reared exclusively on them may suffer from skeletal deformities. It is a good idea to improve the nutritional value of the invertebrates by supplementing their food intake directly. The best method is to use the special supplements or foods now being marketed for this purpose. In the case of mealworms, keep them in a container of chicken meal rather than bran. Bran contains the chemical phytic acid, which in turn combines with available calcium and decreases the nutritional value of the mealworms still further.

Proprietary rearing foods
Apart from livefood, you can now buy a range of proprietary softfoods for breeding birds. These are usually of greater nutritional value than livefoods, but are often ignored by non-domesticated

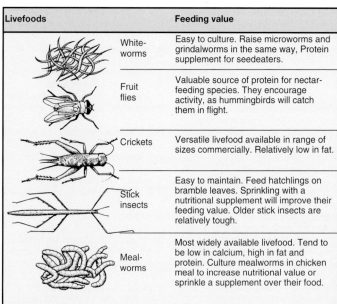

Livefoods		Feeding value
	White-worms	Easy to culture. Raise microworms and grindalworms in the same way. Protein supplement for seedeaters.
	Fruit flies	Valuable source of protein for nectar-feeding species. They encourage activity, as hummingbirds will catch them in flight.
	Crickets	Versatile livefood available in range of sizes commercially. Relatively low in fat.
	Stick insects	Easy to maintain. Feed hatchlings on bramble leaves. Sprinkling with a nutritional supplement will improve their feeding value. Older stick insects are relatively tough.
	Meal-worms	Most widely available livefood. Tend to be low in calcium, high in fat and protein. Culture mealworms in chicken meal to increase nutritional value or sprinkle a supplement over their food.

Above: *A King Parrot feeds on peas. Check foods are fresh.*

Left: *A Grey-headed Parrotbill will relish these berries in its aviary.*

species, since these instinctively seek out livefood when rearing their chicks. However, you may be able to persuade them to sample a food of this type by sprinkling it in a dish containing either livefood or greenfood. With precocial chicks, it is important to offer food particles of the appropriate size, and these are available in the form of starter, or chick, crumbs.

Birds that do not normally take any softfood as part of their regular diet may be tempted to sample such unfamiliar items when they have chicks. Cockatiels, budgerigars and zebra finches are typical examples. Some softfoods can be fed straight from the packet and these are generally more acceptable than those that need to be mixed with water. When mixing food, avoid making it too sloppy, as this decreases its palatability.

Soaked seed
Soaked seed may be essential for some birds, such as cockatoos, that refuse to sample foods other than seed, even through the

How to use	Suitable for
Set up several successive cultures to ensure sufficient supplies.	Small softbills as part of their regular diet, and for finches, especially at breeding time.
Using a specially formulated breeding medium for the culture will avoid unpleasant odours developing. Or breed flies in special feeder with exit hole.	Small softbills as part of regular diet and in even greater quantities at rearing time. Finches relish the wingless variety, which are easier for them to catch.
Cool crickets in refrigerator so they become less active and easier for birds to catch.	Hatchlings are ideal for chicks, finches, and small nectivores. Larger softbills eat adult crickets.
Transfer stick insects from food plant into suitable container to give birds easy access. Do not offer too many, since some may escape around the aviary.	Soft-bodied hatchlings are ideal rearing food for chicks. Taken by finches and softbills.
Sieve out from culture, and transfer to hook-on container for feeding to the birds. They will soon become used to receiving mealworms at a set time each day. Can be used for taming.	Valuable for adult finches and softbills. Chicks may find body casing hard to digest. When the mealworm moults it is more suitable as a rearing food. Also eaten by pheasants.

breeding period. By immersing sufficient seed for one feed in water, you will stimulate the germination process, as well as alter the food value of the seed. Soaked seed has a relatively high protein level – although it still lacks the essential amino-acid residues found in protein of animal origin – and raised Vitamin B levels, making it a valuable addition to a diet of dry seed. Millet sprays and plain canary seed are usually prepared in this way for smaller birds, while black rape is popular with canary breeders. Cockatoos and parrots of a similar size prefer sunflower seed to smaller seeds.

However, once saturated with water, soaked seed becomes a perishable foodstuff. If you supply soaked seed in the morning, it is probably best, especially in the warmer parts of the world, to remove any uneaten remains on the same evening, before they turn mouldy. A few breeders will not use soaked seed for this reason, but if it is prepared carefully it should be quite safe.

Start by washing the seed thoroughly under a running tap to remove dirt and potentially harmful microorganisms, and then immerse it in a container of warm water. Leave it to soak for a day, sieve it and wash it again under a running tap. Finally, drain the soaked seed thoroughly and tip it into a food container for the birds.

Below: *To sprout seeds, rinse and drain twice a day for 4-5 days. Popular for breeding parrots.*

Ringing and record-keeping

As the chicks grow, you will need to decide whether or not to band them with a closed ring. This provides a means of confirming their age and origins, which can be especially important in the case of exhibition stock. Societies often provide their members with special rings that may be required for show birds entered in certain classes at some events. Alternatively, you can buy rings from specialist manufacturers who advertise in avicultural magazines.

Applying closed rings to chicks is not difficult, but it will probably be easier if you can watch a practical demonstration of the technique before attempting it yourself. It is important not to attempt to fit a closed ring too late, otherwise the toes may have grown too large for the band to slide over them and up on to the leg. This depends on the species concerned, of course, but ringing birds at about one week is normally satisfactory.

Start by grouping the three longest toes together, and slide the ring up to the ball of the foot. Keeping the shorter rear toe parallel with the leg, slide the ring up the leg, far enough to release the toe, but without fitting the ring over the knee joint. If it is difficult to free the toe, gently prize it out using a blunt matchstick slid between the toe and leg. Once the toe is free, the ring should slide freely up and down the leg from the foot to the knee joint.

Ringing does entail a certain element of risk, however, both in the short and long term. While most domestic species, such as budgerigars and canaries, accept banding of their chicks, other birds may injure their chicks in attempting to remove the foreign object placed on their offspring's leg. Rubbing peat over the shiny surface of the ring may make it less conspicuous in the nest.

Unfortunately, close-ringing is compulsory for certain species of birds if you intend to sell the offspring at a future date. Check details of the wildlife regulations in your country with the government department concerned. Ring sizes may be specified in some cases.

It is clear that with increasing domestication birds have become larger and their legs thicker. Consequently, rings suitable for wild birds often prove too small for

Below: *Applying a closed ring. Ease the ring over front three toes.*

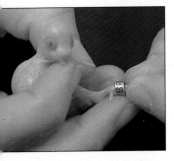

Below: *Fold the fourth toe up the leg and push the ring over it.*

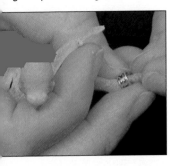

Below: *Release the fourth toe by sliding the ring up the leg.*

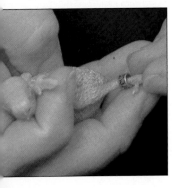

aviary-bred stock. One result of this can be swelling of the leg around the closed ring. This problem can arise even in budgerigars, and prompt veterinary treatment is essential to prevent the blood supply to the foot being cut off, leading to gangrene and eventual loss of the limb. A veterinarian will use a special tool to remove the ring, making the task both easier and safer, since it is not difficult to break the bird's leg.

Research into safer ringing methods is proceeding. When ordering rings, be sure to give the supplier details of the size or species of bird that you are breeding so that you receive the right type of ring for the bird in question. Celluloid rings are suitable for smaller seedeaters and aluminium rings are used for budgerigars. Stainless steel is reserved for the larger parrots, since their powerful beaks can rip through thin aluminium, and they may slice their tongues on the sharp edges of the metal.

Even if you do not show your birds, it may be important to ring them simply for identification purposes. Under these circumstances, it is better to use split celluloid rings. These are available in a wide range of colours and are quite easy to fit, irrespective of the bird's age, since they simply clip over the leg and do not have to be slid over the toes. Before releasing the bird, ensure that the ring is tightly closed, otherwise there is a risk that the bird could become caught up by the ring in the aviary.

In the past, rings of various colours have been used for identification purposes within a group of birds, but in the light of recent research it may be best to restrict the choice of colours. Two-coloured bands, such as orange and green, appear to affect the status and lower the life expectancy of zebra finches that wear them in aviary surroundings.

Even if celluloid rings are not officially coloured and embossed

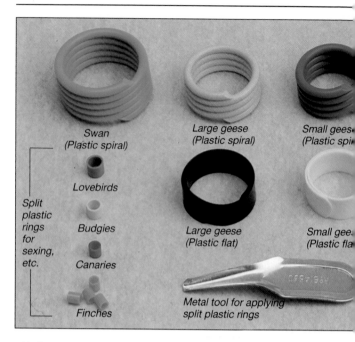

Swan
(Plastic spiral)

Large geese
(Plastic spiral)

Small geese
(Plastic spi

Lovebirds

Split
plastic
rings
for
sexing,
etc.

Budgies

Large geese
(Plastic flat)

Small gee.
(Plastic fla

Canaries

Finches

Metal tool for applying
split plastic rings

with the year date, you can still devise your own system for recognizing the age of birds in the aviary by using a different colour to band the chicks produced each year. You can also use rings to distinguish the sexes of small seedeaters, such as Bengalese Finches, where there is no visual distinction apart from the song of the cock bird during the breeding season. Fit them to the left leg of known cocks, for example, and to the right leg of hens. You should then be able to distinguish both the age of the birds and the sex of the adults, without having to catch them to inspect ring numbers.

Coded information on rings can be a help when you plan your breeding pairs, and accurate record keeping is vital for the progress of an exhibition stud. Start by preparing a nest card for each breeding cage. This shows the ring numbers of the adult birds, and has space to note the date that the first egg was laid and the total number of eggs in the clutch. You can add further details, such as the anticipated date of hatching, the number of chicks,

Above: *Specialist manufacturers produce rings in various sizes. Clearly state the size and type required when you place an order.*

Right: *Accurate record keeping in the form of nest cards and a stock register is vital to the successful management of an exhibition stud.*

the number reared successfully and the ring numbers given to the young birds.

Subsequently, you should transfer this information to a breeding register, where the season's results are recorded, along with individual details concerning the birds themselves, such as show wins. It is then much easier to plan future pairings because you can trace the ancestry of any bird in the stud. Log new acquisitions along with their date of purchase, as well as sales and deaths. Many breeders are beginning to keep these details on personal computer programmes instead of a card index in order to speed the process of selecting breeding pairs of birds each year.

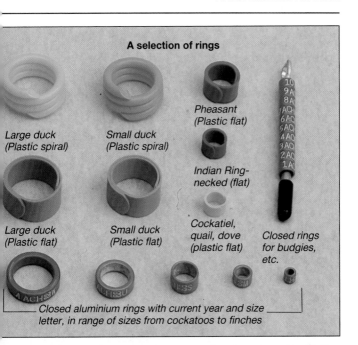

A selection of rings

Large duck (Plastic spiral)

Small duck (Plastic spiral)

Pheasant (Plastic flat)

Indian Ring-necked (flat)

Large duck (Plastic flat)

Small duck (Plastic flat)

Cockatiel, quail, dove (plastic flat)

Closed rings for budgies, etc.

Closed aluminium rings with current year and size letter, in range of sizes from cockatoos to finches

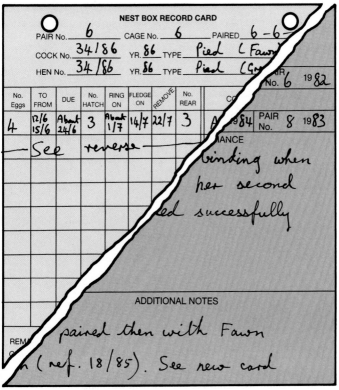

NEST BOX RECORD CARD

PAIR No. __6__   CAGE No. __6__   PAIRED __6-6-__

COCK No. __34/86__   YR. __86__   TYPE __Pied (Fawn__

HEN No. __34/86__   YR. __86__   TYPE __Pied (Gr__

PAIR No. __6__ 19__82__

No. Eggs	TO FROM	DUE	No. HATCH	RING ON	FLEDGE ON	REMOVE	No. REAR	C
4	12/6 15/6	About 24/6	3	About 1/7	14/7	22/7	3	A

PAIR No. __8__ 19__84__   PAIR No. __8__ 19__83__

— See reverse —

binding when her second ed successfully

ADDITIONAL NOTES

paired then with Fawn (ref. 18/85). See new card

57

Fostering

Record keeping is also vital if you foster the eggs or chicks of one pair of birds to another. You may need to adopt this procedure in an emergency, if a hen succumbs with egg-binding, for example. Alternatively, fostering may be carried out as a matter of routine with the eggs of species, such as the Gouldian Finch, partly to improve the reproductive output of these birds. If their eggs are removed soon after laying, the birds should lay again within a short space of time. Furthermore, if the adult birds are known to be poor parents, fostering may give the chicks a greater chance of being reared successfully. The domesticated Barbary Dove is a popular choice as a foster parent for pigeons and doves.

When transferring eggs, try to place them under a pair of birds that laid at about the same time. This will improve the chances of success, since clearly the fostered eggs have little chance of surviving if the other chicks hatch long before them. Try to place the eggs to be fostered under a hen with

Below: *Gouldian and Bicheno chicks in a Bengalese nest. Bengalese are popular foster-parents for Australian finches.*

relatively few eggs of her own, so that she does not find herself with more chicks than she can rear successfully. Before placing it under the new hen, mark each fostered egg with a small cross using a felt-tipped pen. You will then be able to check which of the two sets of eggs have hatched.

It is usually quite easy to foster eggs, since the likelihood of rejection is slight and the chicks are reared in the normal way. If you do decide to foster between species – Australian finches, for example, are often reared by Bengalese finches – be sure to return the young birds to the company of their own species at an early stage. This will ensure that there is no risk of imprinting, which could cause the fostered birds to reject their own species in favour of the birds that reared them.

Fostering chicks is more commonly carried out in an emergency. Once again, there is a greater chance of success if the nestlings are placed alongside birds of a similar age. Synchronization of the breeding cycle is critical, especially in the case of pigeons and doves. These birds produce crop milk, a protein-rich secretion used to sustain the offspring for the first few days of life. It is produced under hormonal

influence as an integral part of the breeding cycle, and cannot be triggered simply by the sudden presence of a chick in the nest. If the chicks of pigeons and doves are to be fostered successfully from a very early age, you must move them to a pair whose offspring are at a similar stage in their development.

Generally, birds do not notice the presence of an additional chick in their brood, but it is useful to distract their attention for a time to allow the newcomer to settle down amongst its nestmates. For example, you can offer greenfood to a breeding pair of budgerigars and then clean the nestbox in the usual way, first adding the new chick to the others in the bowl (see page 49). Transfer them all back together and leave them alone for several hours before checking that all the offspring have been fed. From this point onwards there should be no problems. Younger chicks tend to be accepted much more readily than those that are already several weeks old. This may be because by this stage the

hen is preparing to lay again and is taking less interest in her chicks. At this point the cock bird assumes greater responsibility.

Bantams as foster-parents
The use of foster-parents is not confined to altricial species; broody bantams are traditionally used to hatch the eggs of both waterfowl and pheasants, among others. This does entail extra work, however, compared with using an incubator, as you need to look after the birds throughout the year. If you do opt for a bantam, be sure to choose a reliable fostering strain. Silkie cross bantams – bred from a large Silkie crossed with a smaller bantam, such as a Sussex – are especially popular. If necessary, you can induce broodiness in the foster parent by placing dummy eggs in an open-fronted nestbox lined with straw. The term 'broody' simply describes a hen in breeding

Below: *Bantams used as foster-parents will hatch and brood pheasant and waterfowl chicks.*

condition, and such behaviour is most common during the warmer part of the year. Seeing the uncovered eggs she will be drawn to incubate them.

Once the hen is sitting steadfastly on the dummy eggs, coming off just to feed and drink, you can place the real eggs under her. It is worthwhile dusting her with a special powder or aerosol at this stage, especially around the vent area, to kill mites and lice. These parasites multiply rapidly during the breeding period and transfer to the chicks when they hatch, thus weakening them. Depending on the size of the eggs, it is usually feasible for a Silkie cross to incubate nine eggs quite satisfactorily. Offer her a diet of corn, grit and water throughout the incubation phase.

Rearing fostered eggs
It is possible to determine the fertility of eggs by candling them carefully (see page 66). Discard any that appear clear after they have been incubated for a fortnight. Once the chicks hatch, they should learn to eat quite rapidly from their foster-parent, unlike some incubator-hatched birds, which can be reluctant to feed in the absence of an adult. At this stage certain ducklings, notably those of the Carolina and many teal, will benefit from livefood added to rearing food; the movement of the invertebrates attracts them to the food source and persuades them to start feeding without delay, irrespective of the conditions under which they were hatched.

Ensure that ducklings are kept out of the worst of the weather, housing them and their foster mother in a breeding coop with a run attached. This should be about 120cm(48in) long and 45cm(18in) wide, with the sides of the run covered with mesh.

Locate the run on an area of short grass, so that the chicks are not saturated by a heavy dew and chilled as a result. If necessary, protect them from rain by covering part of the run with translucent plastic. The chicks can move freely in and out of the run, while the broody hen remains in the coop, although she will need regular exercise as well. Provide chick crumbs for the young birds, and shallow water containers that cannot tip over. Plastic-based drinking fountains are quite suitable. As fostered ducklings get older, you can increase the depth of the water and provide a separate container so that their feathers gradually become waterproof and capable of keeping them buoyant, once they take to the water in earnest. Most chicks will grow quite rapidly and you can usually separate them from their foster-parent once they are fully feathered, which will be from about five weeks of age.

Even if you rely on a broody bantam to hatch the chicks it is worth acquiring an infrared lamp (see page 69), should the broody lose interest in her offspring at any stage. This is most likely to occur when they are moved from the nest to the coop and run. Try to minimize the disturbance at this time by transferring the chicks in a cardboard box, and introducing a couple of chicks to the coop first, before placing the broody within. Providing they are not too active, she should soon settle down with them, and the others can be replaced alongside her without delay. Keep them in a warm spot so that they do not become chilled at this stage. It may help to move the groups in the late afternoon when the hen is about to settle down for the night.

Always site the coop and run in a sheltered part of the garden, not exposed to prolonged periods of direct sunlight. You could lose chicks as a result of heat stroke, if they are unable to escape from the sun's hot rays. Overheated chicks will not settle together and appear obviously distressed, gasping with an open beak.

If you suspect heat stroke, be sure to move them to a cooler spot without delay.

Above: *The Muscovy Duck, now available in a domesticated form, can prove a reliable foster-parent.*

Below: *Laying quarters for poultry with a nestbox accessible from outside. Check that it is secure.*

61

Using incubators

For many years incubators were used mainly for hatching the eggs of precocial birds. Their relatively large eggs were less susceptible to slight fluctuations in incubation temperature and the chicks were easy to cater for after hatching. Even a slight variation in temperature could reduce the hatchability of the smaller eggs laid by altricial species. However, the advent of microchip technology has ensured the accuracy and reliability of modern incubators, and has resulted in the manufacture of much smaller models, so that it is quite easy to find space for them in a shaded part of the birdroom or house.

These developments have been accompanied by a much greater understanding of the nutritional needs of the offspring of many species and the formulation of complete diets suitable for rearing parrots and other birds. Together, these factors have ensured that artificial incubation is now carried out more widely and with a greater range of species than ever before. This is clearly of vital importance in the conservation sphere.

Types of incubator

Today, virtually all incubators are operated by electricity, although, in the past, paraffin (kerosene) was often used to supply the heat to warm the eggs. Incubators can be divided into two basic categories: so-called 'still-air' incubators and those that operate on the forced air principle.

Still-air incubators operate quite simply: warm air introduced at the top of the machine cools as it passes down to come into contact with the eggs beneath. Cold air leaves the unit via holes at the bottom. Clearly the temperature may vary considerably at different points within these incubators; indeed, between the top and bottom it may vary by as much as 8°C(18°F). The tray holding the eggs is set about half way down the unit, and it is important to position the thermometer level with the centre of the eggs. This gives

Above: *The Ashy-headed Goose turns its eggs several times a day.*

an average reading allowing for the depth of the eggs, where the constant temperature may vary by only 2°C(5°F) between the top and the bottom. Even so, it is difficult to control the temperature accurately in these incubators.

Ease of operation and reliability have led to an increase in the popularity of forced-air incubators. In this design, the warm air is circulated by a fan at the top of the machine, and does not rely on convection currents alone.

Choosing an incubator

Before you buy an incubator, try to find out as much as possible about the models available. Contact other breeders to discover which incubators they use, and then ask the salesman about the suitability of the model. Select a model that has proved itself effective for the eggs you hope to hatch. This will help to give you an accurate incubation temperature, since slight variations do occur from one design to another. This is because the temperature within the incubator is also linked to factors such as air flow and humidity.

One problem with small incubators, especially those with a clear plastic cover, is a lack of insulation, causing them to lose heat rapidly to their surroundings. It is best to place the machine in a warm environment where heat transference is less likely to pose a serious problem. Kept at an even temperature, there is less need for the heater in the incubator to be constantly controlled by the thermostat.

There is a wide variation in the price of incubators, but hatching just one additional parrot chick will probably justify the increased cost of a more sophisticated machine.

Become familiar with the incubator before you use it in earnest for the first time. It may be helpful, therefore, to visit a dealer to see a model in use. Most specialist suppliers of incubators are very helpful in terms of their after-sales advice and service.

Automatic turning

Where heat loss is likely to be a problem, it is important to select an incubator that features an automatic turning system. It means that you will not have to open the incubator several times a day throughout the incubation period in order to turn the eggs manually. This process, usually carried out by the incubating bird, ensures that the embryo developing within the egg grows normally and does not become fused to the surrounding shell membranes. Regular turning is especially important at the start of the incubation period and failure to carry it out can lead to a high mortality rate closer to hatching.

Hand turning is laborious and increases the risk of jarring the eggs and damaging the developing embryo during the critical early stages of incubation. Furthermore, you may not be available to turn the eggs five or six times every day until they are due to hatch.

Incubator temperature

Before incubation starts, you can store eggs for a short time without any decline in their hatching potential (see page 65). However, once incubation starts, the temperature at which the eggs are incubated clearly exerts a major effect on their potential hatchability. Larger eggs are less at risk from slight temperature fluctuations than are smaller ones. Fluctuations of more than 0.5°C(1°F) can have catastrophic results, with effects being most marked towards the end of the incubation period. If you regularly find that hatchability is poor, then do not overlook temperature variations as a possible cause; you will often find that the thermostat is the culprit.

Below: A forced-air incubator. The optimum water level required across the base will vary according to local conditions. Take humidity readings and adjust the water flow from the reservoir accordingly.

A forced-air incubator

Water reservoir
Air inlet
Power
Fan
Heater coil
Electric motor for automatic turner
Thermostat
Water provides humidity
Dry-bulb thermometer
Wet-bulb thermometer
Air outlet
Turning grid

It is essential to position a reliable thermometer close to the eggs on their centre line. Digital thermometers, although relatively expensive, are becoming increasingly popular, as they are accurate and easy to read from outside the incubator.

Alternatively, you can use a more traditional mercury thermometer. Using two thermometers on opposite sides of the incubator may be better, since you will be able to detect any variations even sooner.

As a basic guide, an incubation temperature of approximately 37.2°C(99°F) will suffice for most species. For parrots, 36.9-37.5°C(98.5-99.5°F) appears to be the maximum range for the incubator setting, if hatchability is not to be jeopardized. In order to measure slight fluctuations in temperature, it is important to use a thermometer with a scale that can show changes of 0.5°C(1°F). For added peace of mind, it may be possible to devise an alarm system, similar to that used on fish tanks, which alerts you to any significant change in temperature.

Incubator humidity
The other environmental factor that plays a vital role in successful hatching is the relative humidity of the air within the incubator. Relative humidity is expressed in terms of the percentage saturation of the air at a given temperature. As the temperature of the air rises, so it is capable of carrying more water vapour. Relative humidity is important during the incubation period because the egg loses water throughout this phase. If the rate of loss is too fast or too slow, this may impair hatchability.

Relative humidity can be measured using a hygrometer, available from garden centres for use in greenhouses. Models of this type with a clock-like face tend not to be highly accurate and, more significantly, are slow to respond to atmospheric changes. However, it might be worthwhile including a hygrometer as a simple means of

Above: *Compare the wet and dry bulb thermometer readings to ensure the correct relative humidity.*

checking the relative humidity from outside the incubator. Do not forget that opening the incubator is likely to affect the relative humidity figure as cooler air enters the unit.

The traditional and most reliable method of monitoring relative humidity is with a wet and dry bulb thermometer. This system relies upon comparing a normal thermometer reading (the dry bulb) with the figure obtained from a second, wet bulb thermometer. The stem of this thermometer is covered for about 3.75cm(1.5in) above the bulb with a wick, usually made of cotton. Immerse the wick in pure distilled water to saturate it. Since the evaporation of water produces a cooling effect, the reading of the wet bulb thermometer will be lower than that of the dry bulb. A comparison of the two figures gives the relative humidity figure in the incubator. Do not use tapwater because mineral salts may crystallize out and affect the water flow and, consequently, the rate of evaporation from the wick. If the wick is not kept fully saturated, the wet bulb will simply give an equivalent figure to that provided by the dry bulb.

A relative humidity reading of about 50% will be adequate to ensure a successful start to hatching parrot eggs, although for species normally found at high altitudes, a lower figure – around 40% – will probably be required. It is possible to manipulate the relative humidity, and thus increase the rate of evaporation from the egg, if it does not appear to be losing sufficient water through the incubation period to permit the formation of an adequate air space (see page 66).

Using an incubator

Before placing any eggs in an incubator, be sure to clean it thoroughly, even if it is new. Strict hygiene precautions are essential, especially between batches of eggs. Since the shell of the egg is porous, any contamination can be fatal to the embryo. This is why dirty eggs are far less likely to hatch than clean ones. Do not place badly contaminated eggs in the incubator, because they will introduce bacteria to the unit. If the shell is soiled, you may be able to remove the worst of the deposit by gently chipping it away. Remember that the egg is obviously fragile and any damage to the shell before the start of the incubation period is likely to prevent it from hatching successfully. You may be able to patch minor shell damage, such as a small shell puncture, with nail varnish (see page 44).

If you hope to hatch the eggs of different species in the incubator (at different times), you will need to buy turning grids of the appropriate size. Those intended for quail eggs will also be suitable for parrot eggs. Generally it is not a good idea to set abnormally small or malformed eggs, since these are unlikely to hatch. It is best to keep eggs of widely different sizes apart, since this can have a detrimental effect on hatchability.

It is not necessary to set eggs as soon as they are laid; indeed, if you wait until a number of eggs have been laid, they can all be set at once and the chicks should start hatching at about the same date. Do not store eggs for longer than a week, however, otherwise their hatchability is almost certain to be reduced. Store the eggs carefully at a constant temperature of about 13°C (55°F), at a relative humidity of 75-85%. The further development of the germinal disc that forms the embryo will not be affected at this temperature. You may turn the eggs once or twice a day during storage, but this is not essential.

Before handling any eggs, wash your hands thoroughly and dry them using clean, disposable paper towelling; a conventional towel may simply spread further bacteria back to your hands. This is especially important before removing eggs from sitting birds. During the normal incubation period, the eggs receive a constant thin coating of feather oil, which contains a substance called lysozyme. This gives the egg some protection against infection, and the oil also helps to prevent the build-up of dirt on the shell.

A recent survey of hatching problems in budgerigar eggs found that two-thirds of the eggs that failed to hatch did so because of bacterial infections during the incubation period. Significantly, staphylococci bacteria were the most common isolates from these eggs, and these are usually present on human hands. Further study revealed that breeders who handled their eggs recorded a higher incidence of death-in-the-shell from this cause than those who left the eggs alone. There was no evidence that the shell was contaminated by dirt in these cases, and typing of the bacteria from the eggs confirmed that many were of known human strains.

Following storage, do not place the eggs straight into the incubator, but allow them to warm up gradually, so that the germ within each is able to adapt to the changing environment. Stand them at room temperature or transfer them to a brooder for several hours. This is particularly important if you already have eggs within the incubator; introducing new, cool eggs will also lower the overall internal temperature and can harm eggs that are already developing.

Candling

You can check on the progress of the eggs during the incubation period using a technique known as 'candling'. This simply entails examining them in a good light.

The name arose in the days before electricity, when gamekeepers used a candle for this purpose. As the embryo develops, it occupies a greater volume of the egg, causing increasing opacity.

You can use a light bulb or even a torch for candling, or you may decide to construct a purpose-built unit in the birdroom, using a 40-watt bulb to illuminate the egg from behind. Be sure to candle an egg as quickly as possible, so there is no risk of a developing embryo becoming chilled, or overheated by the light source. Avoid candling eggs too early during the incubation period, otherwise you may be tempted to discard fertile eggs prematurely.

Above: *Candling is a reliable way of determining whether an egg is fertile. Be sure your hands are clean before handling eggs.*

The air space
The advantage of regular candling throughout the incubation period is that it allows you to keep a check on the development of the air space, which is critical if the chick is to hatch successfully. During the incubation period, the eggs lose water and some breeders regularly weigh them to check progress. Study suggests that under normal circumstances the average water loss from the egg is likely to be about 12-16% of its initial weight. Although it is possible to candle eggs from outside the incubator, by passing a bright torch over the eggs, you will always need to remove them for weighing and thus increase the likelihood of problems arising from handling.

Above: *A newly laid budgerigar egg. At this stage, both fertile and infertile eggs look the same.*

Water loss during the incubation period is a reflection of the relative humidity. If this is too low, then evaporation proceeds too quickly, causing an abnormally large air space to form, and often leaving the chick attached to the shell membranes. Conversely, when the relative humidity is too high, only a small air space develops and this may lead to the chick drowning in the fluid within the egg.

Providing you spot the problems caused by too high a humidity before it is too late, and can isolate the egg in a spare incubator, you can effectively enlarge the air

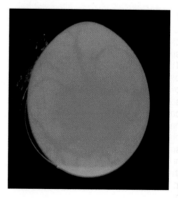

Above: *Seven days after laying, a clear pattern of blood vessels is forming in this fertile egg.*

space by maintaining the egg at a lower humidity, thus drying it out. This is not guaranteed to overcome all problems associated with excessive humidity, however, since, with a smaller air space to sustain it, the chick is likely to hatch at a relatively early stage, with albumen and the remnants of the yolk sac clearly visible. Such a chick is more prone to yolk-sac infections than normal and you must treat its navel without delay. Iodine is useful for this purpose, since it helps to dry up the tissue.

There is often a reluctance on the part of these chicks to feed readily in the immediate post-hatching period, which contributes to a higher mortality rate than normal. The youngsters are often slow to raise their heads for food, and suffer from excessive fluid retention, which gives them a rather bloated appearance, especially around the neck.

Chicks that emerge from eggs incubated under conditions where the relative humidity is too low – a less common problem – are small in size, assuming they survive to this point. The development of their skeletal system is hampered by their inability to extract the maximum amount of calcium from the eggshell, while the kidneys are unable to function effectively towards the end of the incubation period, without sufficient fluid.

Hatching in the incubator

Under normal circumstances, the chick will break through into the air space a day or so before hatching. This is described as internal pipping. As the chick starts to breathe, it uses up the oxygen supply and the rising level of carbon dioxide stimulates hatching. Once you know that the chick has gained access to the air space, it is important to raise the humidity level, in order to prevent the young bird becoming trapped in its shell membranes as they dry out. Avoid unnecessary interference at this stage, since opening the incubator repeatedly will lower the humidity, possibly

jeopardizing hatching.

It is possible to restore the temperature within the incubator much more rapidly than the humidity, which, by definition, lags behind an increase in temperature. Pheasant chicks are especially at risk during this phase because they have particularly tough shell membranes, so raise the relative humidity as high as possible. Keep ventilation to a minimum, since the humidity in the room, averaging 65%, will be far below that needed for successful hatching.

If you have a small number of eggs in a single incubator all at the same stage of development, you may be able to moisten the membranes by painting them carefully, ideally using an artist's paintbrush and distilled water, so that they do not dry out. Obviously, this is difficult to do if the eggs are due to hatch at different times in the same incubator, since they will require different levels of humidity. In this case, set up a second incubator beforehand so that you can transfer eggs which are hatching to conditions of high humidity. Ensure that the temperature is set at a satisfactory level in this second incubator.

You are more likely to encounter problems at the final stage of the incubation process when chicks are being hatched under artificial conditions. These problems are usually caused by mismanagement of the eggs at a previous stage, rather than by any intrinsic genetic weakness in the chicks. Good record keeping will help you to pinpoint when an egg is due to hatch, and this is a useful guide if you need to find out whether an egg is pipping internally at too early a stage, usually because the air space is too small. At the start of the incubation period, number or code the eggs, and keep a written record of their origins and the likely date of hatching. If the chick appears to have difficulty in hatching you will be in a better position to judge if it needs any help (see page 46).

Using a brooder

Once the chick has emerged successfully from the egg, it will be wet and you must take particular care with precocial chicks that they are kept warm and are able to dry off, otherwise they can become fatally chilled. Keep altricial chicks just below the incubator temperature in the immediate post-hatching period, and then transfer them to a brooder. Since they will stay here until they are virtually fully feathered, it is important that the brooder is fitted with a variable temperature control.

Observe the chicks carefully as you alter the temperature. Cold chicks huddle together, showing little enthusiasm to feed, and their crops empty much more slowly than usual. If they are too hot, they will stay as far away from each other as possible and hold their wings out from the body to cool themselves. In extreme cases, they may also pant, with their beaks open. As a general guide following hatching, reduce the temperature over a couple of days to 35°C(95°F) and then lower it further to about 33°C(92°F) until the chicks start to feather up. You can make further downward adjustments as the time for weaning approaches.

Many parrot breeders construct their own brooders. A simple box with ventilation holes and a heat source under reliable thermostatic

Simple box brooder

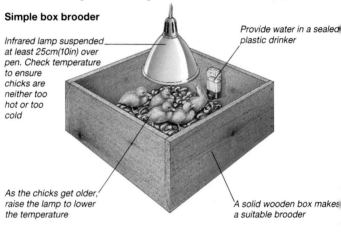

Infrared lamp suspended at least 25cm(10in) over pen. Check temperature to ensure chicks are neither too hot or too cold

Provide water in a sealed plastic drinker

As the chicks get older, raise the lamp to lower the temperature

A solid wooden box makes a suitable brooder

Purpose-built metal brooder

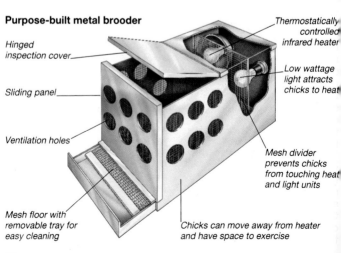

Hinged inspection cover

Sliding panel

Ventilation holes

Mesh floor with removable tray for easy cleaning

Thermostatically controlled infrared heater

Low wattage light attracts chicks to heat

Mesh divider prevents chicks from touching heat and light units

Chicks can move away from heater and have space to exercise

control is adequate. The front usually consists of a sliding panel of transparent plastic, and thermometers monitor the temperature within.

Although ordinary light bulbs are frequently used to heat the brooder, they are not ideal. Firstly, it means that whenever heat is required, either day or night, the lights will be automatically switched on, which may damage the chicks. Secondly, the lifespan of many incandescent bulbs is relatively short when they are suspended at an unusual angle. When a bulb blows, the temperature in the brooder will immediately start to fall. This need not be fatal for the chicks, since you will be attending to them around the clock and will soon detect a fall in temperature, but it can hinder their growth. If you do decide to use incandescent light bulbs, choose two coloured bulbs, such as red or blue, each rated at about 40 watts, which will not be so bright for the chicks' eyes. If one of these cuts out, the temperature fall will be less significant and the risk of chilling is reduced.

Another heating option for the brooder is an infrared heater that emits heat but no light. These are used routinely with precocial chicks. The unit, consisting of the bulb and surrounding reflector, is suspended at a suitable height – at least 25cm (10in) – over the pen containing, say, the ducklings. The heat output is strongest directly under the lamp, with the rays spreading out a variable distance around the unit. The birds can thus adjust their exposure to the heat source in order to maintain an ideal body temperature, moving closer to the unit when they are cold and further away as they warm up.

Place food and an adequate water supply just out of range of the heat source, but within easy reach of the chicks. Always be sure that the birds cannot touch the unit directly. As they grow, you can raise the lamp, which will have the effect of gradually lowering the temperature beneath. This results in a more even distribution of heat across the pen. Always have a spare bulb to hand, otherwise you could find yourself left without a heat source in the brooder when the shops are closed.

Another option, especially as chicks grow older, is a brooder with a heater sealed in its base and a detachable clear acrylic top, which can be cleaned very easily.

Failure to hatch
For a variety of reasons, a proportion of the eggs set in an incubator may fail to hatch. In the first instance, they may not have been fertilized and this becomes evident during the incubation process when you handle them. Poor fertility can result if the cock birds used for breeding are either immature or too old. The age of fertility depends to a large extent on the individual species. Although budgerigars can be fertile at only a few months old, it is likely to be several years before large parrots, such as macaws, can mate successfully. If the pair is not compatible for any reason, even if they are housed under ideal conditions, the hen may lay without mating taking place. A typical example are the *Psittacula* parakeets, such as the Ring-necked (*P. krameri*), which do not form a strong pair bond (see page 33). Alternatively, the egg may have been fertilized, but the embryo died early on, before its development became obvious.

Losses of fertile eggs during the incubation period – be this natural or artificial – can be quite high, and even total under certain circumstances. However, the risk of problems arising is probably higher when the eggs are removed from the birds. Human failings, such as poor storage conditions, excessive handling, dirty surroundings, and inadequate incubator techniques will contribute to losses at this stage. Contamination of the eggs in the nest and chilling can also be contributory factors to losses of parent-incubated eggs.

Hand rearing parrots

One of the most rewarding aspects of breeding parrots is rearing chicks by hand. Watching them develop at close quarters is a fascinating pastime, and will also help the birds to become tame. If you want a parrot as a pet and can hand rear the offspring of a particular pair for this purpose, this is ideal. Be sure to consider the implications, however. Hand rearing is demanding, since newly hatched parrots need feeding about every two hours.

The incubation period provides you with an excellent opportunity to make the necessary preparations for the chicks when they hatch. Even if you are not intending to hand rear young parrots, it is well worth obtaining the basic equipment for use in an emergency. It may be that problems, such as the death of the breeding hen, make this inevitable if the chicks are to survive.

Some breeding farms now offer both incubation and hand rearing services, should you not be in a position to rear the chicks yourself. If you decide to use a hand rearing service, visit the premises in advance so that you can see the conditions under which the chicks will be kept. Establish the terms on which they are accepted, preferably in the form of a written agreement between you and the other party. A breeder will often be happy to accept a percentage of the chicks reared, depending on the number and species concerned. Try to make contact with a local breeder, since this will make it much easier when it comes to moving eggs and chicks. It may not even be necessary for you to carry out the transfer yourself, as collection in a specially heated container may form part of the service.

Preparing food

Following intensive research into the needs of young parrots, especially in the United States, various companies now offer complete rearing foods that are simply mixed with the appropriate volume of water for each feed. All rearing foods must be mixed strictly in accordance with the manufacturer's instructions. These have several advantages over the sometimes complex recipes recommended for this purpose in the past, many of which were based around human infant or convalescent foods. Proprietary rearing foods are carefully formulated to avoid nutritional deficiencies and other possible complications, such as weak bones and retarded growth. The weaning phase can also be easier when using foods of this type.

Hand rearing tools

The easiest and safest way of feeding the chicks is to use a teaspoon with the edges carefully bent in to form a channel for the food. This will enable the chick to take food at its own pace, and lessens the risk of choking it. If food enters the windpipe, inhalation pneumonia may result.

If you look inside the chick's mouth, you will be able to see the opening of the windpipe in the throat, at the back of the tongue. With a sickly weak chick, you may need to feed it directly, taking care to avoid this opening. Ask your veterinarian for a 5ml syringe and a length of suitable plastic tubing for use in an emergency. Regular tube feeding involves passing a volume of food directly into the crop and means there should be no risk of choking the chick.

This method was used quite widely in the early days of hand rearing parrot chicks, but it soon became apparent that it was not entirely satisfactory. Passing the tube into the crop several times a day caused abrasions in the upper part of the digestive tract. These often became infected with *Candida*, a yeast-like microorganism. This infection could spread through the lower part of the digestive tract, with fatal consequences. Candidiasis can also arise with spoon feeding (if hand rearing), but in this case the signs of infection – notably a parallel ridge of whitish growth on the upper mouth where the sides of the spoon have rubbed – are clearly apparent, and are not difficult to treat. Obtain a cream from your veterinarian to rub on to the affected area, and dab the site with a vitamin solution containing Vitamin A to speed recovery. This vitamin, normally stored in the liver, helps to prevent candidiasis.

The feeding routine

It is better to make up a fresh mixture of food on each occasion, although you can store any surplus in a refrigerator for one day and warm it up. You may prefer to divide the food into disposable containers, such as clean, empty yoghurt pots, cut down to a convenient size. Always wash your hands before preparing any food. Use a mixing jug to prepare the required quantity, taking care not to make the food too sloppy by adding too much water. However, if the mixture is too dry, it may become impacted in the crop. Add warm water in small amounts, stirring it into the dry food until you are happy with the consistency. Before feeding the chicks, drop a little food onto the back of your hand to ensure that it is not too hot, or use a thermometer. The temperature should be about 40°C(105°F), although it can be slightly lower for older chicks.

The consistency and quantity of food will alter as the chicks grow. Young chicks need a fairly liquid food and only take a tiny volume.

This is the advantage of dividing the food between smaller containers; having made one batch, it can then be further diluted as necessary. Furthermore, there is less risk of contamination, since there is no longer a communal food container from which all chicks are fed and you can use separate feeding implements.

The frequency of feeding is obviously influenced by the age of the chicks. Younger birds need more frequent feeding – perhaps every hour and a half at first. The crop provides the best means of monitoring their food requirements, and may also show early signs of impending problems (see page 76). The crop is a storage organ at the base of the neck in which food is kept before it passes further down into the digestive tract. In hatchlings, you can plainly see when there is food in the crop; later on, feathers largely obscure this area.

Since you will need to remove the chicks from the brooder for feeding, be sure that the rearing room is kept warm so that the chicks do not lose more heat than necessary. You can also help directly by placing your spare hand around the chick while you feed it. This not only restrains it, but provides body warmth as well.

Remove only one chick at a time for feeding. Place it on a sheet of

Below: *A bent teaspoon is useful for feeding, but check that the edges are not sharp or abrasive.*

Above: *A parrot chick with an empty crop (above) and after a feed (right). Specially formulated diets are simple and reliable.*

clean paper towelling on a firm surface so that you can easily clear up any spilt food. Chicks feed at their own rate and it is not difficult to judge when they have had enough, since you can watch the crop fill up with the food mix. Once the chick appears satisfied, wipe off its beak and replace it in the brooder. At this stage, any food that adheres to the soft tissue of the beak and dries here is likely to cause permanent malformation.

Chicks do better when they are reared in small groups of three or four. The most suitable container for them in early life is a clean, empty margarine tub lined with paper towelling. Some breeders like to use sawdust or wood shavings, but these may be eaten

by the chicks when they become more mobile and cause a fatal blockage in the intestinal tract. Obviously, several chicks huddled together will retain their body heat more efficiently than one on its own, and their growth rate is correspondingly improved. Change the bedding in the container at each feed, so there is no risk of their claws becoming soiled with faecal material.

Obviously you will watch the development of the chicks, but the most reliable means of monitoring their progress is to weigh them at a set time each day, say before the first feed in the morning. You can then be sure of obtaining a reliable set of figures that will reveal any problems without delay. Keep a log, such as a calendar, where you

Below: *Clean plastic tubs, lined with paper towelling, are ideal for housing hand-reared parrot chicks.*

can note the weights of the chicks, and transfer these weights to a graph. Set out the time in days on the horizontal axis and the weight of the individual chicks in grams on the vertical axis. Although this record may not be of particular value with just one clutch of chicks, it can be worthwhile for species that may not have been bred before, or where little previous information has been recorded about their development.

Below: *Typical growth curve for a parrot chick. Slight loss of weight usually occurs at fledging time.*

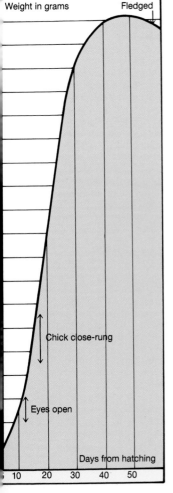

Weight in grams

Fledged

Chick close-rung

Eyes open

Days from hatching

10   20   30   40   50

A graph of the early stages can also provide a baseline for comparing the progress of other chicks later on.

The weaning period

As the chicks grow, they will take increasing quantities of food and you can lengthen the interval between feeds, as long as you ensure that feeding continues on a regular basis. You may need to switch from a spoon to a syringe, in order to cater for the needs of the largest parrot chicks, such as macaws. If in doubt, however, continue with a spoon, replacing the teaspoon with a larger dessertspoon shaped in a similar fashion. As the time for weaning approaches, you will need to introduce more solid food to the chicks' diet. You can mix in ground up sunflower seeds, sold in a hulled form by most health food stores. Some breeders recommend adding a small amount of grit to assist with the digestion of this more solid food.

By the time the chicks are ready for weaning, they will be well-feathered and will need deeper and larger containers to restrain them. From the graph, you will observe that their progress starts to slow down and the curve tends to flatten out. Indeed, weight loss is not uncommon and, having reached a plateau, the graph may dip downwards for a time. This need not normally be a cause for concern, assuming the young bird otherwise appears healthy.

In time, chicks become increasingly restive and are less keen to be fed, often toying with the spoon and consuming little food. At this stage, they need access to a variety of foods that they can sample themselves. Soaked sunflower seed, carefully prepared as described on page 54, is suitable for larger parrots, while millet sprays are favoured by most young birds. You can also offer a variety of pulses such as mung beans. Continue to offer the rearing food, however, reducing the quantity rather than cutting it

73

out suddenly. This could seriously disturb the beneficial bacteria in the digestive system, with potentially fatal consequences.

Weaning should be carried out gradually. By introducing a pelleted diet, equivalent to the rearing food, the process will be more easily accomplished.

Ideally, feed the birds morning and night. Remember that good hygiene is still critical at this stage; remove uneaten food and wash all containers thoroughly. Perches in the rearing cages can become sticky and will need to be scrubbed off or replaced regularly. This applies especially to lories and lorikeets, which are receiving nectar. Never offer this in open pots, but in tubular drinkers, so the chicks are not tempted to bathe in the sticky solution.

It is important to try and keep the plumage of young birds in good condition, unstained by food deposits. It may be useful to hold a piece of paper towelling around the chest area of a fully feathered chick, or to slip a suitably cut out piece over its head. This will keep the worst spillages off the plumage. Lories and lorikeets will appreciate a dish for bathing, whereas other parrots may prefer a light spray with a plant sprayer. Be certain that the chicks do not become saturated or chilled.

General care
Chicks will need increasing space to develop their wing muscles. Initially, they can build up their

Below: *Hand-reared parrot chicks will normally sample a wide range of foods, with beneficial results.*

flying skills by flapping their wings vigorously while perched. They will then attempt to fly, but are often clumsy at this stage. They will not appreciate that there is glass in a window, so be sure to close and screen all windows in the rearing room and thus prevent the chicks flying away or injuring themselves if they escape from the cage. At this stage, it is a good idea to transfer the youngsters to a suitable flight. This should not be too high, so that if the chicks do flop down, it will be quite easy for them to climb back onto a perch.

If you have reared one chick on its own, try to introduce it into the company of other birds of the same species at this stage to avoid permanent imprinting. Kept on its own, the parrot may otherwise relate more to people than to members of its own species. Generally, you should avoid placing the chick directly in a cage with adult birds, but keep it within sight and sound of others of its kind. If the chick is going to another home, wait until it is fully weaned before transferring it to the new owners, unless they are experienced in hand feeding. A change of environment, diet and temperature, even for a short period, may prove fatal. The new owner needs to be aware of these potential dangers.

Hand rearing problems
Each year brings a greater understanding of the problems that can arise when hand rearing chicks, but some losses will always be unavoidable. Indeed, some adult parrots prove bad parents, and the mortality rate under these circumstances may be even higher than with hand rearing. It is, of course, difficult to gain an accurate insight into what may be considered acceptable losses, especially since many breeders are not keen to reveal such figures and individual circumstances have a bearing on the resulting mortality. Do not be disheartened, therefore, when you lose a chick, but obviously try to discover the cause

and prevent it recurring.

Your veterinarian may be of considerable help in this respect. It is quite possible to obtain meaningful post-mortem results, even from young chicks. It is possible to identify problems such as a yolk-sac infection, choking, or fatty liver and kidney syndrome – the latter being a relatively common cause of losses in parrot chicks. As its name suggests, one of the characteristic findings in a case of fatty liver and kidney syndrome is the accumulation of relatively large quantities of fat in the liver. One of the earliest signs of this ailment is a failure of the crop to empty properly, often followed by loss of normal skin colour and rapid death. It may be that some kidney damage occurred before hatching, if the chick came from an egg that became too dry during the incubation period (see page 66).

Dietary factors may also be implicated in the fatty liver and kidney syndrome. Biotin, a member of the Vitamin B group, appears significant; it plays an important role in the synthesis of fatty acids, the 'building blocks' of fat. Egg white contains a compound called avidin which forms an irreversible combination with biotin in the intestines, blocking the absorption of biotin into the body. As a result, you should never use egg white to rear chicks, although the yolk is a valuable source of biotin. Another member of the Vitamin B group, choline also plays a significant part of the metabolism of fat. It can help to counter an abnormal build-up of fat in the liver and elsewhere in the body.

It is possible to supplement the hand-rearing diet with these members of the B vitamin group, and also with the amino-acid methionine, which can be converted in the body to choline. You must take particular care when using supplements with a complete rearing food of any kind for hand rearing purposes, however, since excessive supplementation can be harmful. A more valuable means of countering a specific deficiency in the first instance might be to improve the diet of the adult birds. This would prevent chicks hatching with a relatively low level of biotin, for example. Since many more veterinarians are now involved in the care of parrots, seek advice if you are encountering a high level of fatty liver and kidney syndrome (sometimes abbreviated to FLKS) in your chicks. An analysis of the diet being fed to the young birds may reveal the source of the problem, and allow you to make the necessary adjustments.

FLKS is not the only reason for a failure of the crop to empty normally. Similar signs may result from chilling or the ingestion of foreign material, such as sawdust. It is possible to empty the crop by means of surgery, which sounds drastic, but is in fact a relatively straightforward and safe procedure when carried out by an experienced veterinarian. The site of the incision is stitched as necessary and heals rapidly, with little resulting seepage.

Surgery is safer than trying to empty the crop yourself by massaging the contents of the crop back out of the mouth via the oesophagus. Some of the food may drain into the windpipe. A better first-aid method is to mix up a solution of molasses in water and, after gently rubbing the crop, administer a small volume directly, using a syringe and tubing. Take particular care not to insert the tubing down the windpipe by accident. Following this treatment, it is best to use a fairly liquid diet for the next day or so to flush through the contents of the crop.

A blockage in the crop may lead to a *Candida* infection here. The risk of this disease has been discussed (see page 70), but if it is suspected lower down the intestinal tract, then treatment must begin without delay, since it is a rapid killer of chicks. Raise the Vitamin A level of the diet for two or three days to combat infection.

It is often easier to hand rear parrot chicks that were fed by their parents for a few days after hatching. It may be that the parent birds confer a degree of immunity on their chicks in the early stages of feeding, but this has not really been studied as yet. On the other hand, intestinal worms may be transferred during this early stage, and cause a problem later in the rearing period. Appropriate treatment should resolve such infestations without complications. Always keep such chicks apart from incubator-hatched chicks to prevent any transfer of infection or parasites between the birds.

Skeletal problems have been documented; these affect the legs most commonly, but sometimes also the wings. They are often the result of an inadequate diet during the rearing phase, but each case really needs expert assessment to identify the causal factor or factors. Vitamin or mineral deficiencies, imbalances or even excesses may be responsible. Unfortunately, these problems tend to become most evident towards the end of the rearing period, by which time it may be virtually impossible to correct the deformity. Chicks normally clench their toes when picked up or put

Possible problems during the rearing period

Symptoms/abnormal behaviour	Likely reason
Failure of crop to empty properly; loss of normal skin colour, which becomes very pale	Fatty liver and kidney syndrome (FLKS)
Distended crop, which fails to empty normally	Chilling
	Blockage
White spots or larger whitish areas in the mouth	Candidiasis
Poor growth; refusal to feed	Intestinal worms
Abnormal positioning of legs and wings	Skeletel problems, including fractures, especially in Grey Parrots
Swollen toes	Bacterial infection

down on a surface, so this in itself need not be a cause for concern. If you are worried by the chick's development, seek veterinary advice. Splinting the affected limb or limbs at an early stage may help to correct the problem, especially if the underlying cause has been identified.

Swollen toes by themselves are more likely to indicate an infection. This can arise from a superficial injury linked with dirty nesting conditions. With prompt antibiotic treatment an infection can be resolved, but if left unattended, the affected digit may be lost. Even if treatment is unsuccessful, such parrots can settle well in the home as pets, showing little sign of their

disability. In mild cases, there is no reason why they cannot be retained for breeding stock.

It may be tempting to dispose of most of your hand-reared chicks each year to pet owners, but remember that such birds are invaluable to fellow breeders. Although these parrots will take several years to mature, they will probably nest just as soon as imported birds and will be much better adapted to aviary conditions. In this way, it is possible to establish aviary strains for future generations of birdkeepers, thus helping to perpetuate the species in captivity and possibly to repopulate the wild in the future.

Possible causes	Action required
Kidney damage before hatching, caused by low humidity	Check level of humidity in the incubator
Incorrect diet	Improve diet of parent birds with added Vitamin B especially biotin and choline
Inadequate heat	Improve environmental conditions
Ingestion of foreign material	Administer a solution of molasses and follow with a fairly liquid diet for a day or two. You can massage the crop free yourself, but surgery is relatively safe and may be preferable
Diet low in Vitamin A and/or abrasive feeding techniques	Increase level of Vitamin A, and treat lesions with specific medication from your veterinarian
May be transferred from parents or from other chicks	Keep parent-hatched chicks apart from incubator-hatched chicks to prevent transfer of parasites. Obtain suitable medication from veterinarian for chicks and adult birds. Clean quarters thoroughly
Vitamin or mineral deficiencies, imbalances, or even excesses during the rearing phase. Foot problems may also result from slippery bedding	By the time such problems come to light, it is often too late to effect a remedy. Splinting affected limb(s) at an early age may help to correct the problem. Seek veterinary advice to isolate cause
May arise from superficial injury linked with dirty nest conditions	Prompt treatment with a suitable antibiotic from your veterinarian

The genetics of breeding

A basic knowledge of genetics is vital if you are interested in breeding colour mutations, since it will enable you to pair up your birds in order to produce the greatest number of individuals of a given colour or genetic constitution. Here, we look briefly at the 'language of genetics' and then consider a number of examples that demonstrate genetics in action.

The language of genetics

Genetics is the science of inheritance, reflecting how characteristics are passed from one generation to the next. Its origins date back to the pioneering work of an Austrian monk, Gregor Mendel, in the mid-1800s. Under controlled conditions, he cross-pollinated pea plants of different types and noted the results. From these studies of a number of simple inherited characteristics, he was able to work out how they would be reflected on a statistical basis in succeeding generations. Mendel's work established the fundamental 'laws' of genetics, which are still valid today. Indeed, our clearer understanding of cell biology has merely confirmed the underlying mechanisms that Mendel deduced by purely analytical means.

What Mendel never knew was that within the nucleus of all living cells are a number of threadlike chromosomes on which his so-called 'inherited characteristics', now known as genes, are located. During the process of reproduction, one set of chromosomes (and therefore one set of genes) is transferred from each parent to the offspring. So that the offspring does not have twice as many chromosomes as its parents, the sex cells of each parent, i.e. the sperm and eggs, contain only half the number of chromosomes present in the other cells of the body. Thus, when these two 'half groups' combine at conception (fertilization of the eggs by the sperm) the full complement is restored in the offspring.

It is worth looking at this process in a little more detail. In each 'normal' cell, the chromosomes can be matched up in pairs. During the cell divisions that produce sex cells, these pairs become separated into two matching groups, thus halving the number of chromosomes in each daughter cell. A further cell division splits the two 'arms' of each chromosome into separate cells. Although the genes are largely 'mirrored' on these two arms, they are not necessarily exactly the same. Thus, the genes in each sex cell may not be identical and how they match up with the sex cells of the other parent at fertilization is a totally random affair. In effect, therefore, the genetic 'cards' from both parents are 'reshuffled' in each generation.

Although the offspring receives some genes from each of its parents, this does not mean that it shows a 'mixture' of their characteristics. Quite the opposite. In birds, for example, in which breeders are most concerned about the colour of the plumage, the genes that 'produce' some colours are dominant over others. Such genes are known as 'dominant'; those genes they dominate are called 'recessive'.

A common example will illustrate how this works in practice. If you pair a light green budgerigar with a yellow one, the young will not have both green and yellow plumage. In fact, depending on the genetic make-up of the green budgerigar, all the offspring may very well be green. In budgerigars, the 'green' gene is dominant over the 'yellow' one.

That phrase 'depending on the genetic make-up of the green budgerigar' hints at yet another subtlety in the language of genetics. Quite simply, appearances can be deceiving. The distinctions in question here are the appearance of the bird, the so-called phenotype, and its genetic make-up, or genotype. The green budgerigars that hatch from the eggs of the pairing described above are certainly as green as one of their parents, i.e. they share the same phenotype, but their genetic make-up, genotype, is quite different. The only way of telling one from the other is by further pairings. If the first generation (or F1) green budgerigars are paired together, both green and yellow budgerigars should then appear in the brood.

Just a few more words of explanation will help you to follow the case studies below. Where two identical genes for a certain character occur opposite one another on the paired chromosomes, such as two dominant or two recessive genes, the individual is called 'homozygous' for that character. If the genes are different, i.e. one dominant and one recessive, the individual is 'heterozygous', or 'split', for that character.

Before we look at some examples of genetics in action, we should make the distinction between the two kinds of chromosomes in each cell. In most

animals, one pair of chromosomes determine the sex of the individual. These 'sex chromosomes' are distinct from the other, so-called autosomal, chromosomes (or autosomes) in the cell. An important feature of the sex chromosomes is that one is longer than the other. Thus, some genes on the 'overhang' of the longer chromosome do not have corresponding ones on the shorter one. The significance of this will become clear in the sex-linked mutation case study described later in the section.

Autosomal recessive pairings
Mating a normal green Peach-faced Lovebird with the bluish form, often described as the pastel blue, is an example of a recessive pairing involving the autosomes. The normal green form is dominant over the pastel blue character. In terms of genetic shorthand, the genes of the green form are abbreviated to 'GG', with the use of upper case denoting dominance. The recessive pastel

How sex cells are formed

One pair of 'homologous' chromosomes within the cell nucleus of a normal green/pastel blue (heterozygous) Peach-faced Lovebird

Just before sex cells form, the chromosomes become double stranded

The chromosomes of each homologous pair migrate into separate cells, thus halving their number, and then split into single 'strands' within the sex cells. Of these cells, 50% contain a dominant gene (G) and 50% a recessive (g)

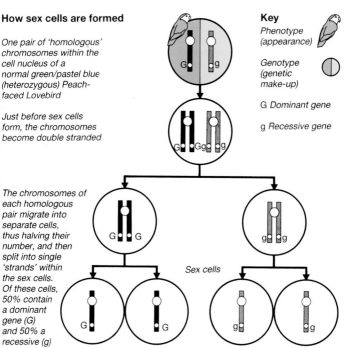

Key

Phenotype (appearance)

Genotype (genetic make-up)

G *Dominant gene*

g *Recessive gene*

Sex cells

79

Here, we continue the genetic sequence started on page 79, in which we saw how the sex cells are formed. Pairing two normal green/ pastel blue heterozygous Peach-faced Lovebirds produces three genetic combinations in the offspring. The 'laws of genetics' predict that they will appear in the proportions shown here. This and the other outcomes of pairings involving recessive genes on autosomes are shown in the panel at the bottom of the page. Box colour shows phenotype.

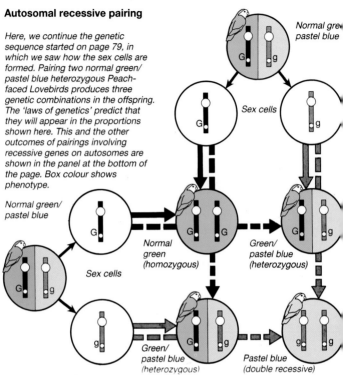

Normal green/ pastel blue

Sex cells

Normal green/ pastel blue

Sex cells

Normal green (homozygous)

Green/ pastel blue (heterozygous)

Green/ pastel blue (heterozygous)

Pastel blue (double recessive)

blue is written as 'gg'. Thus, all the sex cells of the dominant parent will contain one 'G' gene, while those of the recessive parent will contain one 'g' gene. Bearing in mind that all the chicks will receive one of these genes from each parent, it is clear that all the offspring in the first (F1) generation will have a genotype of Gg. This means that phenotypically they are green but they also carry the pastel blue gene in their genotype. The birds are thus heterozygous, or 'split', for the recessive colour of pastel blue.

When these heterozygous birds are paired together, they produce both green and pastel blue chicks in the second (F2) generation. The expected ratio of green to pastel blue chicks will be 3:1, with one third of the green chicks being homozygous (GG) and two thirds being heterozygous green/pastel

Right: All pairings involving the recessive gene for pastel blue.

blue (Gg). The only way of distinguishing between the green and the green/pastel blue chicks is by further pairings since, phenotypically, they will be identical in appearance. The five different possible pairings and the

Normal green (GG)	x	Pastel blue (gg)
Green/pastel blue (Gg)	x	Green/pastel blue (Gg)
Green/pastel blue (Gg)	x	Normal green (GG)
Green/ pastel blue (Gg)	x	Pastel blue (gg)
Pastel blue (gg)	x	Pastel blue (gg)

Above: *A normal green Peach-faced Lovebird. Among the many colour forms, the pastel blue (inset left) is a recessive mutation. This means that the genotype must contain two recessive genes for the colour to appear.*

expected results are shown clearly in the panel below.

The proportions of various offspring in any one nest may not match the predicted results, simply because the combination of genetic material takes place at random. These are the 'average' results, to be anticipated over a large number of matings. This is rather like tossing a coin and calling 'heads' or 'tails'. In just a few throws, you may have significantly more 'tails' than

100% Normal green/pastel blue (Gg)		
50% Normal green/pastel blue (Gg)	25% Normal green (GG)	25% Pastel blue (gg)
50% Normal green/pastel blue (Gg)	50% Normal green (GG)	
50% Normal green/pastel blue (Gg)	50% Pastel blue (gg)	
100% Pastel blue (gg)		

Sex-linked recessive mutation/1

This sequence is the first of two on these pages that follow the sex-linked recessive lutino mutation in Peach-faced Lovebirds. Here, a normal green cock (GG) is paired with a lutino hen (g-). As you can see, no birds of lutino appearance are produced in the first generation. In fact, all the offspring are green, although 50% should be split lutino cocks.

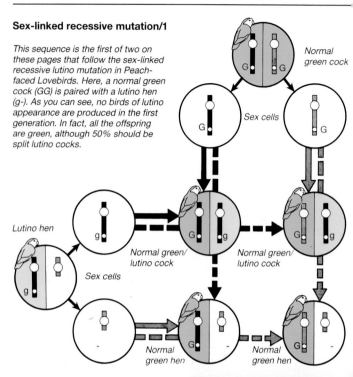

Normal green cock

Sex cells

Lutino hen

Sex cells

Normal green/lutino cock

Normal green/lutino cock

Normal green hen

Normal green hen

'heads', but this will gradually average out as you continue.

All known blue mutations, including those in the Budgerigar and the Ring-necked and Splendid Parakeet, fit into this category. The pied form of the Cockatiel is also a mutation of this type.

Sex-linked recessive mutations
Here, the mutant genes occur on the sex chromosomes. As we have seen, the important point about the sex chromosomes is that in hens one of the pair – the so-called 'Y' chromosome – is shorter than the other, 'Z' chromosome. (The male has two Z chromosomes). If the recessive mutant gene occurs on a part of the Z chromosome that has no counterpart on the shorter Y chromosome, then a hen cannot be split for this particular mutation and her phenotype (appearance) and genotype (genetic make-up) must correspond. In effect, the hen need have only one recessive gene to show the mutation in its appearance, whereas a cock

Sex-linked recessive mutation/2

This sequence shows that it is better to start with a mutant cock (gg) in order to expect some lutino birds in the first generation. In fact, 50% of the offspring should be lutino hens (g-). These can be sexed at an early age by their characteristic red eyes. All the possible pairings involving this particular mutation are clearly shown in the panel on pages 84-5.

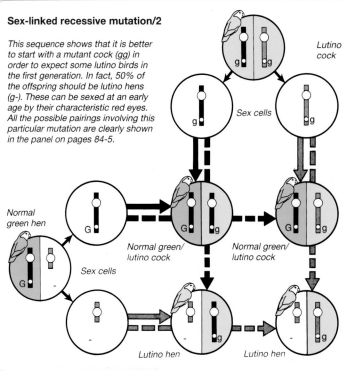

Lutino cock

Sex cells

Normal green hen

Sex cells

Normal green/lutino cock

Normal green/lutino cock

Lutino hen

Lutino hen

Left: *Cinnamon in Peach-faced Lovebirds is a sex-linked recessive mutation. Here, a Cinnamon Pastel Blue with a Cinnamon Light Green.*

needs two recessive genes.

A good example of a sex-linked recessive mutation is the lutino form of the Peach-faced Lovebird. If we use 'G' to represent the dominant gene for the normal green form, then a normal green cock is GG, a normal green hen is G- (where '-' indicates the absence of a corresponding gene), a mutant lutino cock is gg, a split normal/mutant cock is Gg, and a mutant lutino hen is g-. The panel on pages 84-85 shows the results of pairings involving such a sex-linked recessive mutation.

If you wish to breed birds of this type and your financial resources will not stretch to acquiring a pair of mutant sex-linked recessive birds, start with a mutant cock rather than a hen, since you can expect some mutant offspring in the first (F1) generation. (This will not be possible by pairing a mutant hen with a normal cock.) A further advantage of this pairing is that it is possible to sex the chicks with certainty while they are still in the nest, i.e. before their plumage appears: the hens will be lutinos, having distinctive red eyes, contrasting with the dark eyes of the cock birds, whether they are normal green or split lutino.

Most lutino mutations, with the notable exceptions of the rare lutino strain of the Nyasa Lovebird and a form of the lutino Budgerigar that has been lost, fall into the sex-linked recessive category. (The exceptions are autosomal recessive mutations.) Other sex-linked examples include the pearl mutation of the Cockatiel, as well as the cinnamons of several species, and both the fawn and chestnut-flanked white mutations of the Zebra Finch.

Dominant and incomplete dominant mutations

Dominant mutations are effectively the reverse of the autosomal recessive pairing, with the mutant colour form proving dominant to the natural colour form. Instead of being described as 'split' for the natural colour, however, dominance is expressed as being either single factor (sf) or double factor (df). These are equivalent to heterozygous and homozygous.

Continuing with examples drawn from the Peach-faced Lovebird, mating a double factor dominant pied (with genes designated as PP) with a normal green (pp) produces all single factor pied offspring (Pp). It is not possible to distinguish visually between the single and double factor birds, except by further pairings. If any green offspring occur in the pairing of a pied with a normal green, for example, this would prove the pied in question to be single factor. The pairings are shown opposite.

The so-called 'dark factor', identified in the Peach-faced Lovebird, the Budgerigar and more sporadically in other species, such as the Ring-necked Parakeet and

Normal green cock (GG)	x	Lutino hen (g-)
Lutino cock (gg)	x	Normal green hen (G-)
Green/lutino cock (Gg)	x	Normal green hen (G-)
Green/lutino cock (Gg)	x	Lutino hen (g-)
Lutino cock (gg)	x	Lutino hen (g-)

Turquoisine Grass Parakeet, is a dominant mutation. In this particular case, however, it is possible to distinguish between single and double factor birds visually because the latter are darker in coloration. Single factor green birds are described as dark green and double factor birds as olive green. The five possible pairings also involving the normal green, which has no dark factor, are shown in the panel on pages 86-7. In blue series birds, the counterparts of the dark green and olive green are cobalt (sf) and mauve (df).

Dominant pied/df (PP)	x	Normal green (pp)
Dominant pied/sf (Pp)	x	Normal green (pp)
Dominant pied/sf (Pp)	x	Dominant pied/df (PP)
Dominant pied/sf (Pp)	x	Dominant pied/sf (Pp)
Dominant pied/df (PP)	x	Dominant pied/df (PP)

50% Normal green/ lutino cocks (Gg)		50% Normal green hens (G-)	
50% Normal green/ lutino cocks (Gg)		50% Lutino hens (g-)	
25% Green/ lutino cocks (Gg)	25% Green cocks (GG)	25% Lutino hens (g-)	25% Green hens (G-)
25% Green/ lutino cocks (Gg)	25% Lutino cocks (gg)	25% Lutino hens (g-)	25% Green hens (G-)
50% Lutino cocks (gg)		50% Lutino hens (g-)	

Above: *Sex-linked lutino mutation in Peach-faced Lovebirds.*

Crested pairings

In addition to mutations affecting the feather colour, some plumage variations have also arisen, most notably the crested mutation. This has been established in various canaries and the Budgerigar, as well as in Zebra and Bengalese (Society) finches. There is a lethal factor associated with this dominant mutation, however, so that double-factor crested chicks do not survive. As a result, all pairings involving crested birds are made by crossing a single-factor crested bird with a normal one. This produces 50% crested (sf) and 50% normal in the offspring. If you do pair crested (both single factor, of course), the anticipated 25% of the offspring that will be double-factor birds will be rendered non-viable, so the first option is the most effective pairing.

More complex colour forms

Combination of the primary mutations, such as blue and lutino,

Below: *The dominant pied factor in Peach-faced Lovebirds.*

100% Dominant pied/sf (Pp)		
50% Dominant pied/sf (Pp)	50% Normal green (pp)	
50% Dominant pied/sf (Pp)	50% Dominant pied/df (PP)	
50% Dominant pied/sf (Pp)	25% Dominant pied/df (PP)	25% Normal green (pp)
100% Dominant pied/df (PP)		

Normal green	x	Dark green/sf
Normal green	x	Olive green/df
Dark green/sf	x	Olive green/df
Dark green/sf	x	Dark green/sf
Olive green/df	x	Olive green/df

Above: *A single 'dark factor' pro-duces the Dark Green Budgerigar.*

Above: *How the distribution of the dark factor affects plumage colour.*

can give rise to further varieties. Although breeding such colour forms is invariably more complex, it is still possible to predict the likely percentages of the different offspring by arranging all the genetic combinations of both adult birds at right angles to each other.

In order to breed an albino Ring-necked Parakeet, for example, start by breeding a blue hen with a lutino cock, to give lutino/blue hens in the first generation. You

will also need a cock of a similar genotype, obtained by mating a lutino hen with a blue/lutino cock. These can then be mated together as shown in the panel below. Approximately 25% of the offspring will be albinos, in an equal sex ratio, while another 25% will be lutinos. The remaining 50% of chicks will be of the same genotype as their parents, and cannot be distinguished visually from the homozygous lutinos.

A pairing involving two primary mutations (lutino and blue)		Lutino/blue hen (Bbl-) Cocks
	Sex cells	Bl
Lutino/blue cock (Bbll)	Bl	Lutino (BBll)
	Bl	Lutino (BBll)
Considering two primary mutations at the same time, i.e. lutino and blue, makes the picture look more complicated, but the same logic applies. In terms of phenotype (appearance), 25% of the offspring should be albino and 75% lutino. Of the lutino offspring, 25% should be 'pure breeding' in genotype (genetic make-up) and 50% should be 'split', i.e. lutino/blue, to match the genotype of their parents.	bl	Lutino/blue (Bbll)
	bl	Lutino/blue (Bbll)

50% Normal green	50% Dark green/sf	
100% Dark green/sf		
50% Dark green/sf	50% Olive green/df	
50% Dark green/sf	25% Olive green/df	25% Normal green
100% Olive green/df		

Mules and hybrids

Although it is possible to cross closely related species successfully, fertility may be lower under these circumstances because of genetic factors. (If the chromosome numbers do not correspond then breeding cannot occur at all.) A popular pairing of this type involves mating a cock finch, such as a greenfinch, with a hen canary. The offspring, described as mules, show the characteristics of both parents and are highly prized for their song as well as their appearance. (The term 'mule' strictly applies to hybrids bred from specific canary crosses.) Crossing other birds, such as different munias, for example, gives rise to hybrids. Hybrids, generally, are infertile, although a notable exception were the offspring from crossing canaries with the South American Black-hooded Red Siskin (*Carduelis cucullatus*). This pairing laid the foundations for the development of today's red factor canaries.

Cocks	Hens	Hens
bl	B-	b-
Lutino/blue (Bbll)	Lutino (BBl-)	Lutino/blue (Bbl-)
Lutino/blue (Bbll)	Lutino (BBl-)	Lutino/blue (Bbl-)
Albino (bbll)	Lutino/blue (Bbl-)	Albino (bbl-)
Albino (bbll)	Lutino/blue (Bbl-)	Albino (bbl-)

Establishing an exhibition stud

A number of species are now kept and exhibited competitively, being judged on a points scale that relates to the considered 'ideal' for the bird in question. This assessment includes not only the physical appearance (or 'type') of the individual concerned, but also specific details, such as its markings and its condition. Although it may be relatively easy to breed birds of good type, say in the case of the pied Zebra Finch, it is considerably more difficult to produce well-marked individuals. It is harder to exhibit the pied variety successfully than birds with unbroken plumage. You cannot predict with any certainty the variegated markings of the offspring from a particular pair of pied birds, even if they themselves are considered well marked according to the show standard for that particular variety.

Striving for improvement
Developing an exhibition stud of birds of any species will take several years. You need to assess your stock objectively, identifying not only its strengths but also its weaknesses, so that when you pair up birds and buy in new stock, you can begin to rectify any shortcomings. If, for example, you feel that your recessive pied budgerigars are lacking in size, which is not uncommon with this particular variety, do not carry on pairing pied to pied. Instead, obtain a large, good-quality light green budgerigar and pair your best pied stock with this bird. Hopefully, the offspring – having a green phenotype – will introduce an improvement in type, specifically size, when paired with other recessive pieds in your stud.

Line-breeding and in-breeding
Careful record keeping is essential to enable you to trace the origins of your emerging bloodline. Once your birds have reached a good standard, you may decide to refine their appearance yet further by pairing related birds together, instead of using unrelated stock.

Line-breeding entails pairing birds that share a common ancestry, such as two cousins, or a nephew and niece. It is widely practised by many exhibitors, and retains more genetic diversity than in-breeding, where closely related stock is used, such as son being mated with mother or daughter to father.

With in-breeding, the chromosomal variety – and hence genetic variability – is very limited and, unless carried out very carefully, can be self defeating. Apart from emphasizing the strengths of the birds concerned, you also run the risk of highlighting their weaknesses. Reserve in-breeding, therefore, only for the very best birds. Persistent in-breeding over several generations may well give rise to a decline in fertility and an increased incidence of dead-in-the-shell chicks. Make an honest assessment of your stock and, if necessary, introduce fresh blood by out-crossing birds with unrelated stock.

It is usually easier to acquire good-quality cock birds than hens. Indeed, you will find it is more important for the hen to display good maternal instincts when breeding. Therefore, if the bloodline excels in cock birds rather then in hens, you can capitalize on this feature by in-breeding, pairing the cock to his daughter. The resulting progeny must then feature 75% of the cock's genetic input, since their mother received 50% of her genetic base from the same bird. Thus you can maximize on the strengths of an individual bird in your stud within two generations.

If you wish the maternal side to exert a greater influence, you can pair a hen bird with her son. Brother to sister matings are less favoured, since they do not maximize on the existing strengths within the developing bloodline, but simply capitalize on that particular generation's traits.

Although the price asked for exhibition stock is much higher than for ordinary aviary birds, this is a direct reflection of the care

and skill of the exhibition breeder, whose efforts are assessed and acknowledged on the show bench. With a deep pocket, you can 'buy' success, but this will be shortlived unless you are able to develop and maximize the potential of the stock you have acquired. This could be the beginning of a rewarding hobby.

Above: *Line-breeding buff to yellow Norwich canaries. Assess stock for exhibition or breeding by studying your birds closely.*

Below: *This tuneful Roller Canary is a winning exhibit. Select breeding pairs with care for continued show success.*

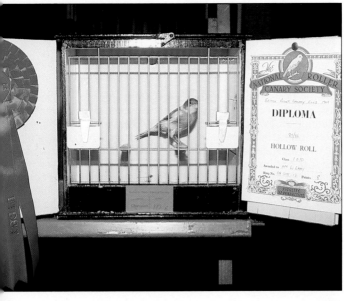

Species section

In Part Two, we examine the breeding habits and requirements of a wide range of birds kept in domestic surroundings. Although the breeding period may vary depending on climate and geographical location, the signs of breeding activity in your birds will be clearly evident. It is well worth making notes on your birds' behaviour at breeding time, partly for your own reference in the future, but also to benefit fellow birdkeepers. A considerable amount of new data is published each year in the various avicultural magazines. Although the first breedings of a particular species tend to attract most attention in the short term, it is the regular breeding of a species over several generations that is likely to be of greater significance. Apart from helping to establish the species in captivity, repeated breeding successes help to give a greater insight into the individual requirements of particular birds, thus improving the quality of their general care.

Co-operation with other breeders is assuming growing importance, especially in the field of parrot breeding; the high cost of acquiring and housing many of the larger psittacines may deter you from buying a group of birds on your own. Working with someone else who shares your interest in a particular species benefits both of you, and your expenditure is effectively halved. Rather than obtain four pairs, for example, you can each decide to buy two, and then hopefully exchange youngsters in the future, ensuring that there is unrelated stock available for use beyond the F2 generation. Alternatively, you may join forces with a fellow enthusiast who owns, say, an unpaired softbill. In this way, there is hope of the birds breeding, rather than living permanently on their own. Loan schemes for the purposes of breeding birds are now becoming more common; contact your local society or look in the birdkeeping magazines for details.

FINCHES

Finches are often described as 'seedeaters', but many species feed on a much wider range of foodstuffs, particularly when they are breeding. The provision of suitable livefood is often vital if the chicks are to be reared successfully in aviary surroundings.

Zebra Finch

Poephila guttata

● **Sexing:** Hens usually lack the chest and flank markings of cocks and have paler beaks.
● **Clutch size:** Up to 6 eggs.
● **Incubation period:** 12 days.
● **Rearing period:** Chicks start to leave the nest about 18 days after hatching and will be eating independently within a fortnight.
● **Diet:** Livefood is not necessary during the rearing phase, but adults usually take softfood.

The Zebra Finch has proved to be the most free breeding of the Australian grassfinches, and a number of popular mutations are now established.

Once they have laid, remove any remaining nesting material so that the birds cannot continue building the nest and bury the eggs in the process. They may also try using pieces of greenfood for nest building. Chop it into small pieces

Above: **Zebra Finch**
A cock bird with zebra-like throat and chestnut flank pattern.

to discourage this behaviour and to prevent the eggs becoming obscured or the greenfood turning mouldy in the nest.

Gouldian Finch

Chloebia gouldiae

● **Sexing:** Hens can be distinguished by their paler breasts. When cock birds are in breeding condition, their beaks turn reddish at the tip, whereas those of the hens turn darker overall.
● **Clutch size:** Up to 6 eggs.
● **Incubation period:** About 16 days.
● **Rearing period:** Chicks leave the nest about 21 days after hatching.
● **Diet:** Soaked seed is important during the rearing phase, along with greenfood and softfoods.

The most striking of the Australian species, this finch naturally occurs with red, black or yellow (orange) head coloration. New colours include a white-breasted form.

In Europe, Gouldian Finches are usually bred in cages, and they will use either a domed nesting basket or an open-fronted nestbox. Eggs are often fostered to Bengalese, but Gouldians should be encouraged to rear their own chicks as well, on a second clutch, for example. Weaning needs to be carried out carefully, as losses among young birds, which are much duller than the adults, can be high. Like the Greenfinch, they often suffer from the condition known as 'going light' (see page 95). Avoid introducing the chicks suddenly to a diet of dry seed, or weaning them before you are certain that they are fully able to feed themselves, to lessen the risk of such problems arising.

Bengalese Finch
Lonchura domestica
● **Sexing:** This species cannot be sexed visually, but cocks in breeding condition may be distinguished by their song.
● **Clutch size:** In a few cases, a hen may lay up to 9 eggs in a clutch.
● **Incubation period:** 15 days.
● **Rearing period:** Chicks emerge from the nest when they are 21 days old.
● **Diet:** Provide eggfood and similar items for rearing purposes.

Below: **Black-headed Gouldian**
May nest in winter. Provide heat.

This finch of hybrid origin does not occur in the wild. The Bengalese makes an ideal introduction to the hobby of breeding birds, the only problem being to sex the individuals correctly. A variety of colour forms are now established, including Self Chocolate, Chestnut and Fawn, all of which have pied counterparts showing variable degrees of white plumage. There is also a crested variety.

Below: **Self Fawn Bengalese**
A mutation with no white feathers.

Java Sparrow
Lonchura oryzivora
● **Sexing:** The song of the cock is really the only means of distinguishing between the sexes. However, cocks may develop an enlarged lower mandible when in breeding condition.
● **Clutch size:** 4-6 eggs.
● **Incubation period:** The eggs are incubated by both parents for 14 days.
● **Rearing period:** Chicks emerge from the nest after 28 days and will be independent within a further 21 days.
● **Diet:** As with other related species, livefood is not essential for rearing purposes.

A member of the Nun or Mannikin group, these birds are more likely to breed when kept on a colony system, with an adequate choice of nesting sites. They will lay their clutch of 4 to 6 eggs in a budgerigar nestbox. Their claws can become overgrown, and may damage the eggs unless they are cut back at the start of the breeding season. In addition to the white (pied) form, there is an attractive fawn variety which originated in Australia but is now quite common in Europe. For photograph see page 29.

Canary
Serinus domesticus
● **Sexing:** No clear visual distinction, although cocks sing during the breeding period, whereas hens tend to cheep.
● **Clutch size:** 4 eggs.
● **Incubation period:** The hen is often left on her own to incubate the eggs, which takes about 14 days.
● **Rearing period:** 21 days.
● **Diet:** Softfood is vital for rearing purposes; also offer other food items, such as greenfood.

This is another widely known domesticated species. The canary is thought to date back perhaps four centuries or more; its ancestors probably came from the islands off the northwest coast of Africa. Since then, distinctive breeds have been developed, often with strong regional links, such as the Norwich in England, which was developed around the county town of Norfolk. At present, the Fife Fancy – derived from the Border Fancy – is rapidly growing in popularity among birdkeepers.

Below: **Canary**
The closed leg ring confirms the bird's age and breeding details.

Greenfinch
Carduelis chloris
● **Sexing:** Hens are generally duller overall.
● **Clutch size:** 3-8 eggs.
● **Incubation period:** About 14 days.
● **Rearing period:** About 15 days.
● **Diet:** A British Finch Mixture or a seed mixture consisting of small cereal seeds and oil seeds, such as rape, plus berries in season and suitable greenfood.

This finch is native to much of Europe and also extends into parts of Asia. A number of colour mutations, including a striking lutino, have been evolved in captive-bred greenfinches and these may be subject to ringing requirements in countries such as the UK, where they are also found in the wild.

Weaning is often a difficult phase for the chicks; they are prone to losing weight across the breastbone – described as 'going light' – and often die, even following antibiotic treatment. Gradual weaning may help to reduce the incidence of 'going light', which is not a specific disease, but can result from a number of causes, such as the bacterial disease, pseudotuberculosis. Clean surroundings and slow weaning, as described for the Gouldian Finch (see page 92) may help. Post-mortem examinations of affected birds may enable you to identify the problem and, hopefully, avoid any such losses with future clutches of chicks.

Orange-cheeked Waxbill

Estrilda melpoda

● **Sexing:** Adult hens are usually slightly paler in coloration than cocks and may have a small area of orange coloration at the sides of the head.
● **Clutch size:** 4-6 eggs.
● **Incubation period:** About 12 days.
● **Rearing period:** The young birds leave the nest at about 21 days old.
● **Diet:** Although waxbills eat seed for much of the year, they usually become highly insectivorous when breeding. Chicks are unlikely to be reared successfully without an adequate supply of livefood.

Above: **Greenfinch**
Watch chicks closely at weaning.

This species tends to be nervous, and will nest more readily in aviary than in cage surroundings. Keep the birds in a planted flight and do not interfere when they start breeding, otherwise they may desert their nest. In common with other related species, the Orange-cheeked Waxbill's nest is a relatively bulky structure. An interesting feature is the so-called 'cock's nest' above, which is relatively conspicuous, yet never occupied. This serves to confuse potential predators, since the real entrance to the breeding chamber beneath is hidden on the side of the nest and often disguised with a feather, carefully positioned.

Below: **Orange-cheeked Waxbill**
Livefood is vital for the chicks.

Golden Song Sparrow
Passer luteus
● **Sexing:** Cocks have yellow bodies and chestnut brown wings, while hens are typically sparrow-like in appearance.
● **Clutch size:** 3-4 eggs.
● **Incubation period:** 12 days.
● **Rearing period:** Chicks emerge from the nest when only a fortnight old.
● **Diet:** Livefood is less essential for rearing this species than it is for waxbills. Softfood and soaked seed are also good rearing foods.

It is easy to detect the onset of breeding condition in these African birds; the beak of the cock changes in colour from pale horn to dark black. They breed better as a group, preferring to build their own nests, lined with feathers, in a suitable clump of gorse thickets hung in the aviary. In the wild, they often seek out thorn bushes at this time, for protection against any potential predators.

Orange Bishop
Euplectes orix
● **Sexing:** Cocks in colour have characteristic markings. Out of the breeding period the cock is much duller, resembling the hen, while still retaining odd black feathers over the body.
● **Clutch size:** 3 eggs.
● **Incubation period:** 14 days.
● **Rearing period:** Chicks fledge at about 14 days old.
● **Diet:** A mixture of cereal seeds, plus some greenstuff and livefood.

Above: **Golden Song Sparrows**
The duller hen is shown on the right. The black beak of the cock indicates breeding condition.

Below: **Orange Bishop**
A cock bird in breeding colour. Keep several hens to each cock.

A member of the group known as weaver birds, because of the cock's habit of weaving a number of nests from strands of dried grass and other material. For breeding purposes, house several hens with one cock; these birds are polygamous and one hen on her own is likely to be heavily persecuted. When the chicks fledge – at about a fortnight old – watch out for any signs of persecution by the cock and, if necessary, remove him from the aviary. All the work of incubation and rearing will be undertaken by the hen bird.

Fischer's Whydah
Vidua fischeri
● **Sexing:** Cocks can be distinguished by their long tail plumes during the breeding period, when they are described as I.F.C. (In Full Colour). They resemble hens when Out Of Colour (O.O.C.).
● **Clutch size:** 3 eggs.
● **Incubation period:** As for the host species, the Purple Grenadier Waxbill, namely 12 days.
● **Rearing period:** About 20 days.
● **Diet:** A mixture of small cereal seeds such as millet, plus some livefood and greenstuff.

While some whydahs undertake the hatching and rearing of their own chicks, this species is entirely parasitic. After mating, hens lay their eggs in the nest of the host species, which in this case is the Purple Grenadier Waxbill (*Uraeginthus ianthinogaster*). Here their chicks hatch, and grow up alongside those of the waxbills, even mimicking their inner mouth markings to confuse the adult birds. Many whydahs will lay their eggs only in the nests of a specific host, but the Pin-tailed Whydah (*Vidua macroura*) has been recorded as using 19 different species as foster parents for its eggs. In aviary surroundings, both the whydah and the host species must obviously be nesting simultaneously if breeding is to be successful. Keep them together in well-planted aviaries.

Above: **Fischer's Whydah**
A cock bird in full colour. In eclipse plumage, cocks resemble hens.

97

PARROTS
These birds are well known for their powers of mimicry. Although in the past, they were kept on diets consisting of little more than dry seed, many species will benefit greatly from a more varied diet that includes pulses, greenfood and fruit. Many breeders are now using special pelleted diets that provide a much more balanced diet than seed alone, although parrots may be reluctant to sample them at first.

Budgerigar

Melopsittacus undulatus
● **Sexing:** Sexing is by cere colour; that of the cock is blue or purplish, while the adult hen's cere is brownish in colour, most noticeably when in breeding condition.
● **Clutch size:** 4-6 eggs.
● **Incubation period:** 18 days.
● **Rearing period:** Young budgerigars will usually emerge from the nest at about 35 days old and will be independent in a further 7 days.
● **Diet:** Offer food supplements, such as packeted softfoods and an iodine 'nibble'. Soaked seed, particularly millet spray, is also a popular rearing food.

The budgerigar has been bred in very many colours and combinations since it was first brought to Europe in 1840 and is now the best-known member of the parrot family. One of the reasons underlying the budgerigar's rise in popularity is its ability to thrive and breed on a diet consisting of little more than dried cereal (grass) seeds. It is probably the easiest member of the parrot family to breed, either in cage or aviary surroundings, although a pair may refuse to nest on their own, since they are social, even when breeding.

Turquoisine Grass Parakeet

Neophema pulchella
● **Sexing:** Cocks can be distinguished by their red wing patches.
● **Clutch size:** The hen will lay about 5 eggs in an average clutch, although double this number is not unknown.
● **Incubation period:** 18 days.

Above: **Budgerigar**
Yellow-faced Cinnamon Cobalt.

Below:
Turquoisine Grass Parakeet

● **Rearing period:** The chicks grow quite quickly, sometimes fledge before 28 days, and are soon independent.

● **Diet:** A parakeet seed mixture, consisting largely of small cereal seeds, with some sunflower seed.

These attractive parakeets are popular aviary occupants and generally breed quite freely. Watch for signs of aggression from the cock, however, especially if the pair have not bred before. As with most other parrots, keep pairs of birds apart during the breeding period. Breeding details are similar to those of the budgerigar, and pairs can produce two rounds of chicks in a season. They are prone to intestinal worms, and it is a good idea to treat breeding birds regularly with a suitable proprietary remedy and to clean the floor of their quarters thoroughly before each breeding season.

Cockatiel

Nymphicus hollandicus

● **Sexing:** Hens are duller overall. Adult hens show barring on the undersurface of the tail feathers. Difficult to sex visually before first moult at about 6 months.

● **Clutch size:** The average clutch consists of about 5 eggs, but larger numbers are not unknown.

● **Incubation period:** Eggs are incubated by both parents for about 19 days.

● **Rearing period:** Young birds leave the nest about 35 days after hatching. Feather-plucking is a vice particularly associated with cockatiels of the lutino mutation. It can be a reflection of the adults' desire to breed again; the cock, especially, may pluck the chicks in an attempt to persuade them to leave the nest before they are ready to do so. If they are plucked, take care to ensure that they roost under cover.

● **Diet:** Bread and milk, or a high protein insectivorous food, may be readily accepted by a breeding pair of cockatiels.

Cockatiels are also quite free-breeding, and a number of colour forms have been developed, especially during recent years. The pale yellow lutino is perhaps the most striking. These birds can breed at a year old, and may live well into their twenties. Withdraw the nestbox to prevent them laying during the winter when chicks will be vulnerable to the cold.

Below: **Cockatiel**
A pearl cinnamon (left) and a pied (right). Cockatiels are docile and quiet, as well as free-breeding.

Umbrella Cockatoo
Cacatua alba
● **Sexing:** These large cockatoos can usually be sexed on the basis of their eye coloration; hens have reddish-brown irises, whereas those of cocks are black.
● **Clutch size:** 2 eggs.
● **Incubation period:** 25 days.
● **Rearing period:** About 80 days.
● **Diet:** A parrot seed mixture, supplemented with pulses, parrot pellets, greenfood and fruit.

Try to select a pair that appear to be bonded, as this may help to prevent displays of aggression at the start of the breeding period. Incubation of the eggs is shared by the parents. Be sure to provide a stout nestbox to deter the adult birds' beaks. Arrange easy inspection, so that you can remove the chicks without delay if you suspect they are being neglected.

Right: **Umbrella Cockatoo**
A cock bird, as is evident from the black irises. These noisy birds can be aggressive at breeding time.

Red Lory
Eos bornea
● **Sexing:** Difficult to sex visually. However, cocks generally have broader heads and beaks, and may appear larger overall.
● **Clutch size:** 2 eggs.
● **Incubation period:** 24 days.
● **Rearing period:** 63 days.
● **Diet:** Nectar mixture, fruit and a little seed.

Lories and lorikeets originate from Australasia, but are quite hardy once acclimatized. Pairs of Red Lories have nested successfully indoors, but do not be misled into thinking that because two birds spend long periods of time preening each other, they are of the opposite sex. However, true pairs often prove prolific and reliable breeders.

Right: **Red Lory**
A free-breeding and attractive bird. Some pairs will pluck their chicks before they leave the nest.

Vernal Hanging Parrot
Loriculus vernalis
● **Sexing:** Hens usually lack the blue throat markings and have brown irises.
● **Clutch size:** Up to 4 eggs.
● **Incubation period:** About 21 days.
● **Rearing period:** The chicks will leave the nest about 28 days after hatching.
● **Diet:** In addition to nectar and other regular foods, the birds may accept a rearing food and even livefood, such as mealworms as extra protein for their chicks.

These small parrots have earned their common name from their habit of roosting upside down, hanging from branches, where their plumage blends effectively with the background of leaves. When breeding, hanging parrots line their nesting chamber with leaves that they carry tucked into their body plumage. Other members of the parrot family lay their eggs on alternate days, but hens of this species lay every day.

Below: **Vernal Hanging Parrot**
Hens lack blue throat markings.

Ring-necked Parakeet
Psittacula krameri
- **Sexing:** Hens lack the characteristic neck collar of the cock bird.
- **Clutch size:** 6 eggs.
- **Incubation period:** 23 days.
- **Rearing period:** 49 days.
- **Diet:** Parrot seed mixture, millet sprays, pulses, greenfood and fruit, especially sweet apple.

This species has the widest distribution of any psittacine, occurring in both Africa and Asia. The lutino is the most common colour mutation, but the blue and albino – the result of pairing primary lutino and blue mutations – (see *The genetics of breeding*, page 78) are also becoming more numerous. Other colours include cinnamon and grey. These parakeets often nest early in the year and so, although they are generally quite hardy, you should encourage them to roost under cover to avoid any risk of frostbite.

Grey Parrot
Psittacus erithacus
- **Sexing:** Hard to distinguish, but mature cocks may have darker wings.
- **Clutch size:** 3-4 eggs.
- **Incubation period:** 28 days.
- **Rearing period:** Chicks will be ready to leave the nest about 84 days after hatching.
- **Diet:** Parrot seed and pellets, pulses, greenfood and fruit.

These African parrots have been kept as pets for centuries, but only quite recently have they been bred in captivity in any numbers. They may not attempt to breed until at least their fourth year, but once they start breeding they have a long reproductive life and may continue breeding into their thirties. However, they are often poor parents, and chicks that hatch may need hand rearing.

Right: **Grey Parrot**
Each year, more of these birds are hatched in aviaries. Hand rearing is time consuming, but rewarding.

Left: **Ring-necked Parakeet**
A cock bird, showing the neck ring.
New colour forms are also popular.

Senegal Parrot

Poicephalus senegalus
● **Sexing:** Cannot be sexed
visually.
● **Clutch size:** 3-4 eggs.
● **Incubation period:** 28 days.
● **Rearing period:** 63 days.
● **Diet:** Parrot seed and pellets,
plus greenfood, pulses and fruit.
They are especially fond of
groundnuts (peanuts).

Adult Senegals tend to be nervous
by nature, and this can handicap

their breeding attempts. They may
also show a preference for nesting
during the winter months, when
the likelihood of failure is
significantly increased. In order to
encourage breeding activity, site
the nestbox in a quiet and fairly
dark part of the aviary, rather than
in bright light. The risk of egg-
binding will be minimized if you
place the nestbox in the shelter.
Hens in breeding condition often
flare their tails around the nestbox.

Below: **Senegal Parrots**
Pairs can only be recognized by
surgical sexing. They can become
quite destructive when breeding.

Right: **Peach-faced Lovebird**
Unlike other parrots, lovebirds collect nesting material to line their nest. Popular and attractive.

Peach-faced Lovebird
Agapornis roseicollis
● **Sexing:** Like the Senegal, these African parrots cannot be sexed visually.
● **Clutch size:** About 5 eggs.
● **Incubation period:** 28 days.
● **Rearing period:** Chicks fledge about 42 days after hatching.
● **Diet:** A mixture of the smaller cereal seeds, with some sunflower seeds, plus greenfood and fruit.

A whole host of colour forms of this species have been bred in recent years, including the pastel blue, which is sea-green in colour, yellow, white and pied forms, to name but a few. Hens are largely responsible for carrying nesting materials to the box. If both birds undertake this task, and ten or so clear (that is, infertile) eggs result, it is almost certain that they are both hens. Be careful when introducing lovebirds together – despite their name, these birds can be savage towards each other.

Below: **Celestial Parrotlet**
These small parrots from South America mature early and breed successfully inn flights or cages.

Celestial Parrotlet
Forpus coelestis
● **Sexing:** Hens are duller than cocks, being largely green with just a bluish tinge to their rumps.
● **Clutch size:** Up to 6 eggs.
● **Incubation period:** 18 days.
● **Rearing period:** About 30 days.
● **Diet:** Smaller cereal seeds, including millet sprays, plus greenfood and fruit.

Unlike the larger parrots, which take several years to mature, parrotlets can breed when only six months old. A compatible pair should prove quite prolific, nesting repeatedly. However, these small parrots can also be highly aggressive towards each other, and even their chicks. Remove the young birds without delay if the cock bird shows any signs of aggression towards them, or they may be attacked.

Golden-crowned Conure
Aratinga aurea
● **Sexing:** Cannot be sexed visually.
● **Clutch size:** Up to 4 eggs.
● **Incubation period:** About 26 days.
● **Rearing period:** Chicks fledge at around 49 days of age and are independent several weeks later.
● **Diet:** Brown bread soaked in milk is an ideal rearing food.

These birds, also described as Peach-fronted Conures, are less noisy than some *Aratinga* species. Place nestboxes in a secluded part of the aviary and they will nest quite readily. Young Golden-crowned Conures can be recognized by the clear band of yellow plumage alongside the orange markings on the forehead. In addition, their beaks are paler than those of the adults.

Below: **Golden-capped Conure**
One of this vivid group of conures.

White-fronted Amazon
Amazona albifrons
● **Sexing:** Cocks have red rather than green wing coverts. However, it appears from surgical sexing that just a few apparent cocks, with red wing coverts, are in fact hens.
● **Clutch size:** Up to 4 eggs.
● **Incubation period:** 28 days.
● **Rearing period:** 60 days.
● **Diet:** Parrot food and pellets, plus fruit, pulses and greenfood.

It may take several years for young birds to mature. The onset of breeding condition can be detected by observing their behaviour. At this time, the parrots will be noisier, and you are likely to see tail flaring. Hens incubate the eggs alone, but both adult birds are likely to be aggressive towards anyone approaching the aviary during the breeding period. Under normal circumstances out of doors, Amazons will lay only once a year, but indoors they may nest twice in quite rapid succession. Birds kept in an outside flight may not lay until early summer in northern temperate climates.

Above: **White-fronted Amazon**
A hen, with green wing coverts.

Below: **Blue and Gold Macaw**
True pairs can nest for many years.

Blue and Gold Macaw
Ara ararauna
● **Sexing:** No visual distinction is really possible, although cocks may have larger, bolder heads.
● **Clutch size:** 2-3 eggs.
● **Incubation period:** About 30 days.
● **Rearing period:** The chicks remain in the nest longer than other species and will be about 120 days old before they leave.
● **Diet:** Parrot food and pellets, plus nuts, such as brazils, greenfood, pulses and fruit.

These giants of the parrot family, nearly 90cm (36in) long, need sturdy nesting facilities; a wooden barrel would be ideal. They may not start breeding until four or five years old, but pairs usually prove consistent once they start nesting. When hand rearing large macaw chicks, watch out for skeletal problems which may result from their rapid growth.

SOFTBILLS
Seed forms a very insignificant part of the diet of these birds. They feed on other items, such as fruit, nectar and insects, depending on the species. Livefood assumes greater importance in the diet of all softbills during the breeding period when they have chicks in the nest.

Purple Glossy Starling
Lamprotornis chalybeus
● **Sexing:** Cannot be sexed visually.
● **Clutch size:** 2-3 eggs.
● **Incubation period:** 14 days.
● **Rearing period:** The young birds will leave the nest at about 21 days and will be independent within a further week.
● **Diet:** Softbill mixture or pellets, fruit and invertebrates such as mealworms.

Starlings and mynahs are perhaps one of the easiest groups of softbills to breed satisfactorily in aviary surroundings, partly because of their omnivorous feeding habits. However, sexing is a problem. Starlings may become highly aggressive when breeding. Ensure that the hen is not being unduly persecuted and remove the chicks as soon as possible after they leave the nest; they may be attacked and even killed, if the adults are keen to breed again.

Below: **Hartlaub's Touraco**
House these large birds in individual pairs when breeding.

Above: **Purple Glossy Starling**
Provide a nestbox for breeding.

Hartlaub's Touraco
Touraco hartlaubi
● **Sexing:** No visual distinction possible between the sexes.
● **Clutch size:** 2-3 eggs.
● **Incubation period:** Incubation is shared and lasts about 30 days.
● **Rearing period:** The chicks will often venture from the nesting platform when only 21 days old, but it will be several more weeks before they are independent.
● **Diet:** Softbill mixture, fruit, chopped greenfood, such as spinach, and livefood.

Unlike starlings, these birds do not use a nestbox; instead, they construct a thin platform of twigs on which they lay their eggs. Like starlings, the cock can be very spiteful towards his intended mate. Ensure that the aviary is densely planted, to enable the hen to retreat if she is being harassed by the cock bird. In severe cases, you may need to clip one of the cock's wings. Be sure to monitor introductions closely and separate the individuals if necessary.

Levaillant's Barbet

Trachyphonus vaillantii
- **Sexing:** Hens tend to be duller in overall coloration than cocks.
- **Clutch size:** 3-5 eggs.
- **Incubation period:** About 14 days.
- **Rearing period:** Chicks emerge just over 21 days after hatching.
- **Diet:** These birds often become highly insectivorous when they are in breeding condition.

An African member of a family found in both the Old and New World (i.e. in Africa, Asia and the Americas.) The birds use their stout beaks to enlarge existing cavities in trees, and placing bark around the entrance hole of a nestbox may stimulate breeding activity. Some barbets will attempt to tunnel into the aviary floor when breeding, and you will need to ensure that they are not at risk from flooding when it rains.

Above: **Levaillant's Barbet**
This tree-nesting African species needs ample livefood for breeding.

Red-billed Hornbill

Tockus erythrorhynchus
- **Sexing:** Hens generally have smaller, paler beaks.
- **Clutch:** Up to 6 eggs.
- **Incubation period:** 28 days.
- **Rearing period:** 42 days.
- **Diet:** Invertebrates are ideal for rearing purposes; supply them in large quantities.

Although most hornbills are too large for the average collection, this species is sometimes available and its breeding habits make it a fascinating addition to the aviary. The female is actually sealed inside the breeding chamber, leaving just a small hole through which she is fed by the male. In aviary surroundings you will need to provide a clay mixture for this purpose, that the birds will dilute with their droppings. The hen will break out 10 weeks later, to be followed by her offspring. These birds are not entirely hardy and are susceptible to frostbite.

Right: **Red-billed Hornbill**
The hen, sealed in the nest, is fed by the cock until she breaks out.

PIGEONS AND DOVES

There is no clear distinction between the descriptions of 'pigeon' and 'dove', but generally, pigeons tend to be the larger members of the family. Many species can be prolific under aviary conditions, although some may be reluctant to incubate their eggs. Fostering is preferable to artificial incubation and hand rearing in these circumstances.

Above: **Diamond Dove**
These birds are the popular normal grey form.

Above: **Barbary Dove**
Less nervous than most larger doves. Breed in cage or aviary.

Diamond Dove
Geopelia cuneata
● **Sexing:** Both sexes have red eye rings but those of the cock become much more swollen in breeding condition. Cocks also have more spots on the wings than hen birds.
● **Clutch size:** 2 eggs.
● **Incubation period:** 13 days.
● **Rearing period:** Chicks may leave the nest as early as 11 days after hatching. At this stage they will be unable to fly properly, however, and may become saturated on the floor of the flight.
● **Diet:** small cereal seeds.

This species provides an ideal introduction to breeding doves, since pairs invariably prove diligent in their parental duties. Both adults will sit on a canary nestpan where they will take it in turns to incubate the eggs. Various colour mutations, such as silver, cream and red, which is a russet shade, are now established in this species.

Barbary Dove
Streptopelia risoria
● **Sexing:** Hens are darker in coloration.
● **Clutch size:** 2 eggs.
● **Incubation period:** 14 days.
● **Rearing period:** The young leave the nest for the first time at around 17 days old.
● **Diet:** The young doves are fed initially on crop milk, which has a high level of protein. Adult birds will frequently rear their chicks without additional foodstuffs, but a suitable softfood will be beneficial.

These are gentle and steady domesticated doves, widely used as foster parents for less reliable breeding birds. They will even nest successfully on the floor of the cage, and will perch freely on the hand, being naturally quite tame.

109

PHEASANTS AND QUAIL
These birds tend to be polygamous when breeding. House one cock with several hens, as males can be aggressive, especially at breeding time. If necessary, you can hatch the eggs quite easily in an incubator, and rearing the chicks is not generally difficult.

Chinese Painted Quail
Excalifactoria chinensis
● **Sexing:** Hens are easily recognized by their brownish overall coloration, lacking the bluish plumage of the cock.
● **Clutch size:** About 6 eggs.
● **Incubation period:** 18 days.
● **Rearing period:** There is no rearing period as such with this group or with waterfowl (pages 47-48); chicks are capable of living on their own virtually from hatching providing they are kept in a favourable temperature. They are not fed by their parents.
● **Diet:** Small cereal seeds, plus some livefood.

These small ground-dwelling birds need to be kept in groups with one cock to several hens. A single hen will almost inevitably be persecuted and may lose feathers, especially around the neck, and possibly sustain more serious injuries. In addition, the constant attention of the cock tends to distract the hen from incubation duties. As domestication has proceeded, these quail have become more reluctant to incubate their own eggs, and you may need to hatch them artificially. Keep the tiny chicks carefully confined to prevent them escaping or succumbing in wet weather.

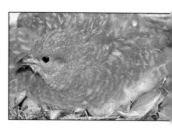

Above: **Chinese Painted Quail**
Provide cover on the aviary floor.

Golden Pheasant
Chrysolophus pictus
● **Sexing:** Hens are much duller than cocks. It can take two years for cock birds to attain full colour.
● **Clutch size:** Up to 12 eggs.
● **Incubation period:** Around 21 days.
● **Rearing period:** Like other pheasants and quails (see page 47), chicks are virtually independent at hatching.
● **Diet:** A pheasant seed mix.

House several hens with one male bird. Introduce the cock separately to each hen for several days, and then remove him after mating has taken place. The hen will lay and, hopefully, hatch the chicks on her own. Provide suitable rearing food.

Below: **Golden Pheasant**
Magnificent and inexpensive.

Indian Blue Peafowl
Pavo cristatus
● **Sexing:** Males can be distinguished by their magnificent coloration and full tail plumes, which they achieve when they are about three years old.
● **Clutch size:** Up to 6 eggs.
● **Incubation period:** 30 days.
● **Rearing period:** 60 days.
● **Diet:** Larger cereal seeds, such as wheat and maize, and items such as soaked dog meal, greenfood and invertebrates.

The magnificent display of cock birds of this polygamous species is well known. Hens choose a secluded spot where they incubate the eggs alone. They then remain with their chicks, brooding them for as long as two months after hatching. Several mutations have been recorded, including both white and pied forms.

Below: **Indian Blue Peafowl**
This is an adult cock bird, displaying its superb tail plumes.

WATERFOWL
A number of species of ducks, geese and swans can be kept in the surroundings of an average garden. Geese, especially, do not need a large area of water, although they and swans are more aggressive than ducks. A number of species are relatively dull outside the breeding period, when the males are in eclipse plumage, and sexing can prove more problematical at this stage.

Mandarin Duck
Aix galericulata
● **Sexing:** Drakes are more colourful than ducks. When out of colour, in eclipse plumage, males can be distinguished by their yellow feet and reddish beaks, those of ducks are greenish and grey respectively.
● **Clutch size:** 12 eggs.
● **Incubation period:** 28 days.
● **Rearing period:** Chicks are fairly independent at hatching.

● **Diet:** Typical duck food. Duckweed is a popular rearing food, and eggfood may also be taken during the breeding period.

These attractive ducks are tree nesters in the wild, and need a raised nestbox, complete with an access ramp, to encourage them to nest in captivity (see page 16). The pair bond is very strong and the birds are not aggressive. They are hardy, colourful and easy to cater for. Justifiably popular.

Above: **Mandarin Duck**
A drake, in breeding plumage.

Carolina Duck

Aix sponsa
● **Sexing:** Drakes are more colourful than ducks in breeding condition. Out of colour, the drake can be distinguished by his reddish beak.
● **Clutch size:** About 12 eggs. May lay two clutches in succession.
● **Incubation period:** 28 days.
● **Rearing period:** Chicks are virtually independent on hatching.

● **Diet:** Typical duck diet, including cereal seeds and possibly pellets.

The Carolina Duck is similar in its habits to the Mandarin Duck, but tends to prove more prolific than its relative, usually laying two broods during the breeding season. A chromosomal distinction prevents hybridization of these two species if they are kept together.

Below: **Carolina Duck**
A colourful, tree-nesting species.

Barnacle Goose

Branta leucopsis
● **Sexing:** Ganders are generally larger than geese.
● **Clutch size:** 6 eggs.
● **Incubation period:** 26 days.
● **Rearing period:** Chicks are independent almost immediately after hatching.
● **Diet:** A duck or goose mix that includes grain. Will also browse on grass and other greenfood.

Geese are more terrestrial than ducks, and will benefit from access to an area of short-cut lawn where they can graze. The pair bond is strong and the gander will protect the sitting female, driving away intruders into the breeding territory by hissing vigorously.

Below: **Barnacle Goose**
Keep these in a group on their own to prevent hybridization.

Black Swan

Cygnus atratus
● **Sexing:** Hens (known as pens) are smaller, with shorter necks.
● **Clutch size:** Up to 6 eggs.
● **Incubation period:** 36 days.
● **Rearing period:** Chicks are virtually independent on hatching.
● **Diet:** Duck or goose mix. Will also browse on greenstuff.

Majestic but aggressive, swans need a relatively large area of water. Keep a pair on their own; the male (cob) will attack both rivals and, often, other waterfowl, especially during the breeding period. They construct a bulky nest, often partially concealed in a reed bed, and share the incubation duties. The adults will protect the cygnets in the water, even carrying them on their backs. As with all waterfowl, it is wise to curb their flight, either permanently by pinioning at a very early age, or by regularly clipping the new flight feathers during the moult. Otherwise the young birds may take off to find a new territory as they approach independence.

Below: **Black Swan**
These will form a strong pair bond.

Index

Page numbers in **bold** indicate major references, including accompanying photographs. Page numbers in *italics* indicate captions to other illustrations. Less important text entries are shown in normal type.

Picture credits

Artists
Copyright of the artwork illustrations on the pages following the artists' names is the property of Salamander Books Ltd.

Rod Ferring: 52; John Francis: 43, 45, 63, 68; Alan Harris: 35; Maltings Partnership: 16(part), 23; Clifford and Wendy Meadway: 51; David Noble: 79, 80, 82, 83; Guy Troughton: 14-15, 26, 27, 36, 38-39, 45; Peter Young: 16(part)

Photographs
Unless otherwise stated, all the photographs have been taken by and are the copyright of Cyril Laubscher. The publishers wish to thank the following photographers who have supplied other photographs for this book. The photographs have been credited by page number and position on the page: (B)Bottom, (T)Top, (C)Centre, (BL)Bottom left etc.

Aquila Photographics: 59(E.A. Janes), 61(T, E.A. Janes)
Bruce Coleman: 47(Jane Burton)
Ian Hunt © Salamander Books: 32, 54, 56-57
Ideas into Print: 18(T, D.Brown), 44(D. Brown), 50(D. Brown), 55(D. Brown), 61(B)

Acknowledgements
The publishers wish to thank the following for their help in preparing this book: A.C. Hughes; Robin Haigh Incubators; Paul & June Bailey; Sharon Bailey; Lionel Ball; Eric Barlow; Christine Baxter; Bob Beeson; Bentley Wildfowl Trust; Blean Bird Park; Peter Clear; Dave & Rose Coles; Marion & Andrew Cripps; Alan Donnelly; Ken & Shirley Epps; Ray & Kathleen Fisk; Sharon Hurrell; Tim Kemp; Alan Jones; George Lewsey; Louis & Heather Martin; Stanley Maughan; Oaklands Park Farm Aviaries; Ron Oxley; Brian Pettit; Mike & Jane Pickering; Mick & Beryl Plose; Porter's Cage Bird Appliances; Janet & Alan Ralph; Ron Rayner; Val Read; Dave & Sandra Rowe; Fred Sherman; Stan & Jill Sindel; Charlie & Jane Smith; George Smith; Steve Stephenson; Nigel Taboney; Mick & Jean Uden; Joyce Venner; Cliff Wright.

Further reading

Alderton, D. *The Complete Cage and Aviary Bird Handbook*, Pelham Books, 1986
Alderton, D. *A Birdkeeper's Guide to Softbills*, Salamander Books, 1987
Alderton, D. *A Birdkeeper's Guide to Finches*, Salamander Books, 1988
Alderton, D. *A Birdkeeper's Guide to Budgies*, Salamander Books, 1988
Anderson-Brown, A.F. *The Incubation Book*, World Pheasant Association, 1985
Arnall, L. & Keymer, I.F. *Bird Diseases*, Bailliere Tindall, 1975
Burr, E.W. (Ed.) *Companion Bird Medicine*, Iowa State University Press, 1986
Forshaw, J.M. & Cooper, W. *Parrots of the World*, David and Charles, 1978
Goodwin, D. *Estrildid Finches of the World*, British Museum (Natural History), 1982
Goodwin, D. *Pigeons and Doves of the World*, British Museum (Natural History), 1983
Harper, D. *Pet Birds for Home and Garden*, Salamander Books, 1986
Howman, K.C.R. *Pheasants – Their Breeding and Management*, K & R Books, 1979
Immelman, K. *Australian Finches*, Angus and Robertson, 1982
Kay, D. *Bantams*, David and Charles, 1983
Kolbe, H. *Ornamental Waterfowl*, Gresham Books, 1979
Low, R. *Parrots – Their Care and Breeding*, Blandford Press, 1986
Low, R. *Hand-rearing Parrots and Other Birds*, Blandford Press, 1987
Mobbs, A. *Gouldian Finches*, Nimrod Book Services, 1985
Robbins, G.E.S. *Quail – Their Breeding and Management*, World Pheasant Association, 1984
Trollope, J. *The Care and Breeding of Seed-eating Birds*, Blandford Press, 1983
Walker, G.B.R. *Coloured Canaries*, Blandford Press, 1976
Walker, G.B.R. & Avon, D. *Coloured, Type and Song Canaries*, Blandford Press, 1987
Walters, J. & Parker, M. *Keeping Ducks, Geese and Turkeys*, Pelham Books, 1976
Watmough, W. & Rogers, C.H. *The Cult of the Budgerigar*, Nimrod Book Services, 1984
Wheeler, H.G. *Exhibition and Flying Pigeons*, Spur Publications, 1978

Grey Peacock Pheasant hen incubating eggs